MODERN RUSSIAN MILITARY AIRCRAFT

MODERN RUSSIAN MILITARY AIRCRAFT

RYAN CUNNINGHAM

amber BOOKS

First published in 2022

Reprinted in 2025

Published by Amber Books Ltd
United House
London N7 9DP
United Kingdom
www.amberbooks.co.uk
Facebook: amberbooks
Instagram: amberbooksltd
Twitter: @amberbooks
Pinterest: amberbooksltd

ISBN: 978-1-83886-201-5

Editor: Michael Spilling
Picture research: Terry Forshaw

Printed in China

Contents

Introduction

Russia possesses the second largest fleet of military aircraft in the world, after the United States. The majority of these serve in the Russian Air Force, since August 2015 one of the two component parts of the Russian Aerospace Forces – the other being the Russian Aerospace Defence Force, concerned primarily with anti-aircraft artillery and missile defence. A smaller number of aircraft are also operated by the Russian Navy.

The origins of Russian military aviation lie in the Tsarist period. A separate air arm of the army was created in 1912 and in World War II, when the Imperial Russian Air Service fielded the world's first four-engine strategic bomber. After the October Revolution of 1917 the Workers and Peasants Air Fleet evolved into the Soviet Air Force, which weathered the onslaught of Nazi Germany's Operation Barbarossa in 1941 to become one of the most powerful air forces by 1945. Throughout the Cold War, Soviet aviation generally matched that of the US in both capability and numbers: in 1980, for example, it was the largest air force in the world with around 10,000 aircraft. But the calamitous economic collapse caused by the dissolution of the Soviet Union in December 1991 resulted in a drastic curtailment of funding. Crew training, aircraft development and infrastructure all suffered and the ability to obtain basic items such as fuel, or even

Emblematic of modern Russian combat rotorcraft, the Kamov Ka-52 Alligator combat helicopter possesses remarkable levels of agility, as demonstrated here by '51 Yellow' at the International Aviation and Space Salon (MAKS) in Zhukovsky, Moscow.

The latest combat type to enter Russian service, Su-57 deliveries have taken place at a trickle since the first production aircraft entered service in December 2020. The first regiment of the Su-57 is expected to be equipped with its full complement of 24 aircraft by 2025. '051', pictured, is the first flying prototype.

pay their personnel, became highly problematic for the Russian state. The Russian air arm is still dealing with the repercussions of this financial strain today.

New types

Since 2000, the situation has consistently improved and it is notable that several stalled aircraft programmes have restarted since the turn of the century, such as the Mil Mi-28 'Havoc' and Ka-52 'Hoodlum' attack helicopters, and the Su-35 'Super Flanker'. Other types have seen production restart after a pause of more than a decade, most dramatically the Tu-160 'Blackjack'.

New types have also appeared, the most exciting being the Su-57 'Felon', the capabilities of which, particularly in comparison to its most obvious equivalent, the F-22 Raptor, continue to be the subject of intense speculation and fierce debate. Despite costing a fraction of the F-22, the Su-57 is still colossally expensive and production aircraft are appearing at an extremely slow rate.

Like the US Air Force, virtually all modern Russian military aircraft in service today are developed versions of late Cold War types.

Combat service

In contrast to their US counterparts, Russian, and indeed Soviet, aircraft have seen relatively little active service in Russian hands. The Soviet Union's operations over Afghanistan saw the initial use of many of the familiar types in service today. Since 1992, Russian aviation assets have been operationally deployed in the two Chechen Wars, the 2008 South Ossetia War against Georgia and, from 2015 onwards, in an ongoing involvement in the Syrian Civil War. More recently, the Russian Aerospace Forces have been committed since February 2022 to their toughest test yet, the invasion of neighbouring Ukraine. At the time of writing it is unclear how effective they have proved, with several aircraft lost and air supremacy not yet achieved against the Ukrainian Air Force, themselves equipped with many of the same aircraft types.

The criteria which aircraft had to meet to be included in this book are that they are currently in service with the Russian Aerospace Forces or Navy and that the type was either designed or manufactured in Russia – thus some Ukrainian designed types, such as the An-124, are included as these aircraft have been built in Russian factories.

FIGHTERS AND ATTACK AIRCRAFT

Despite containing some other important and effective types, Russia's fighter force continues to be dominated, as it has been for the last 30 years, by derivatives of the Sukhoi Su-27 'Flanker' and Mikoyan MiG-29 'Fulcrum'. Both aircraft have been subject to various redesigns and upgrades to maintain their capability, and both remain formidable.

This chapter includes the following aircraft:

- Mikoyan MiG-29
- Mikoyan MiG-29K
- Mikoyan MiG-29M and MiG-35
- Mikoyan MiG-31
- Sukhoi Su-24M
- Sukhoi Su-25
- Sukhoi Su-27
- Sukhoi Su-30
- Sukhoi Su-33
- Sukhoi Su-34
- Sukhoi Su-35
- Sukhoi Su-57
- Sukhoi Checkmate

Various iterations of Sukhoi's hugely successful 'Flanker' still form the backbone of Russia's fighter fleet. Condensation forms over the leading edges of '901', first prototype of the Su-35BM, as it shows off the type's spectacular manoeuvrability at the 2011 MAKS airshow.

Mikoyan MiG-29 NATO Reporting name: Fulcrum

Originally developed in 1971 as a heavy fighter similar to the Sukhoi Su-27, with which it shares its overall configuration. The MiG-29 was optimized for air defence, securing local air superiority as a broad successor to the MiG-21.

The Soviet Union decided to pursue separate heavy and 'light' fighter projects, although more accurately the MiG-29 is a medium-weight design. Soviet-operated aircraft in the Cold War era also had a tactical nuclear-strike role. Only more recently has the design been adapted to provide a true multirole capability. The first flight of the MiG-29 occurred on 6 October 1977, and after a batch of 14 development aircraft the first pre-series aircraft took to the air in early 1979, followed by initial production aircraft in the summer of 1982.

Becoming operational at Kubinka in 1980 and first operationally based outside the Soviet Union in 1986 at Wittstock, East Germany, the MiG-29 was only officially commissioned into service in 1987 after various improvements had been made. The original production run was 620 of the initial single-seat 9.12 Fulcrum-A, the early production version. The 9.12A was slightly downgraded for Warsaw Pact export, while 9.12B was for non-Warsaw Pact export. In total, 528 examples of the improved single-seat 9.13 Fulcrum-C version were constructed, featuring additional internal fuel capacity as well as provision for two 1150-litre (253-gallon) drop tanks under the wings (added to the standard 1500-litre/330-gallon auxiliary fuel tank that can be carried under the fuselage between the engines) and a new Gardenia electronic jammer in the bulged fuselage spine.

Weaponry

The basic armament of the MiG-29 consisted of two medium-range R-27R (AA-10 Alamo) air-to-air missiles with semi-active radar guidance on inner pylons, plus four short-range R-73 (AA-11 Archer) or R-60M (AA-8 Aphid) air-to-air missiles with infrared guidance on the outer pylons. As an

MiG-29 Fulcrum
Weight (maximum take-off): 19,700kg (43,431lb)
Dimensions: Length 16.28m (53ft 5in), Wingspan 11.41m (37ft 5in), Height 4.73m (15ft 6in)
Powerplant: Two Klimov/St Petersburg RD-33 (*izdeliye* 88) turbofans each rated at 49.42kN (11,100lbf) dry, and 81.4kN (18,300lbf) with afterburning
Maximum speed: 2400km/h (1491mph)
Range: 2100km (1305 miles) with under-fuselage drop tank
Ceiling: 18,000m (59,050ft)
Crew: 1
Armament: One 30mm (1.2in) Gryazev-Shipunov GSh-30-1 autocannon in the port leading-edge root extension; up to 4000kg (8800lb) of weapons and stores

MiG-29BM

A product of the ARZ-558 aircraft repair plant in Baranovichi, Belarus, the MiG-29BM upgrade is similar to the MiG-29SM, with N019P, R-27ER and R-77 AAMs, plus precision air-to-ground munitions, cockpit tweaks, and inflight refuelling probe. Eight have been upgraded.

Slovakia's MiG-29 fleet was upgraded to MiG-29AS standard from 2005 onwards. This example was photographed taxiing at the SIAF airshow and clearly shows the aircraft's FOD guards to prevent objects being ingested into the air intakes when running on the ground.

alternative to the air defence stores, up to 4000kg (8818lb) of offensive stores can be carried on the same six pylons, including tactical nuclear bombs, freefall conventional bombs, napalm tanks, sub-munitions dispensers and various unguided rockets. All versions have a GSh-30-1 single-barrel 30mm (1.2in) cannon in the port leading-edge root extension with 150 rounds.

In addition, 197 two-seat MiG-29UB Fulcrum-B versions were built with an additional seat in front of the pilot and a slightly extended fuselage. On two seaters the radar was deleted, but IRST, cannon and weapons pylons were retained. The last version to appear in the Soviet era was the MiG-29S (*izdeliye* 9.13S) advanced version of the Fulcrum-C with modernized N019M Topaz radar of increased range, plus R-77, R-27T and R-27RE/TE compatibility. The MiG-29S flew in 1990 but only a small batch was produced for domestic service. Large-scale production ceased in 1993 after 1345 production aircraft were built. However, with so many incomplete aircraft left at the factories, small-scale production continued, as well as smaller numbers of newly-built aircraft.

Export development

Much development has taken place in the post-Soviet era beginning with the MiG-29SD export derivative of the Fulcrum-A and MiG-29SE export derivative of the Fulcrum-C with N019ME radar, R-77 missile and some Western navigation and communication systems with deliveries to countries including Eritrea, Myanmar, Peru and Sudan. The MiG-29SM (*izdeliye* 9.13SM) upgraded export version

MiG-29

The Phazotron-NIIR N019 radar has a Cassegrain antenna; the search range is 70km (43.5 miles) in a head-on position and 35km (21.75 miles) in a tail-on position against a fighter-size target, with an increase of about 50 per cent for larger, bomber-size targets. The radar can track-while-scan up to 10 targets simultaneously and engage the one chosen by the computer as the most important.

Nose
The early MiG-29's nose contained a N019 Phazotron pulse-doppler radar with a relatively short accurate range (60km) when compared to its Western contemporaries. A disadvantage offset to some extent by the provision of IRST (Infra-red Search and Track).

Control surfaces
The leading edge of the wing is fitted with full-span manoeuvering flaps that are computer controlled. The trailing edge is outfitted with plain flaps inboard and ailerons outboard.

MiG-29 Fulcrum

Weight (maximum take-off): 19,700kg (43,431lb)
Dimensions: Length 16.28m (53ft 5in), Wingspan 11.41m (37ft 5in), Height 4.73m (15ft 6in)
Powerplant: Two Klimov/St Petersburg RD-33 (*izdeliye* 88) turbofans each rated at 49.42kN (11,100lbf) dry, and 81.4kN (18,300lbf) with afterburning
Maximum speed: 2400km/h (1491mph)
Range: 2100km (1305 miles) with under-fuselage drop tank
Ceiling: 18,000m (59,050ft)
Crew: 1
Armament: One 30mm (1.2in) Gryazev-Shipunov GSh-30-1 autocannon in the port leading-edge root extension; up to 4000kg (8800lb) of weapons and stores

Tail
The twin canted tailfins are combined with all-moving horizontal tail surfaces.
Unlike Western counterparts, the early MiG-29s relied on mainly conventional, hydraulically-operated flying controls.

Powerplant
In its basic form, the MiG-29 is powered by a pair of RD-33 turbofans, offering increased thrust in afterburning when compared to Western engines, albeit with slightly reduced dry output.

with N019MP radar and air-to-ground capability was provided to Peru as an upgrade for its aircraft, producing the MiG-29SM-P that also features a refuelling probe.

The MiG-29N (and two-seat MiG-29UBN) is an export version for Malaysia with a refuelling probe.

Prototype version

A long-range version of the MiG-29SM upgraded export aircraft was conceived by the Russkaya Avionika company during the mid-1990s, initially as the *izdeliye* 9.17. It featured avionics based around a Mil Std 1553B data bus and N019MP radar and included air-to-ground weapons, glass cockpit with two digital displays and HOTAS controls. A large 2020-litre (444-gallon) fuel tank was fitted in the fuselage spine, plus refuelling probe. Only a prototype was built but work on the aircraft was continued by MiG as the MiG-29SMT, now with N010M radar and different avionics.

The large spine fuel tank on the 9.17 was aerodynamically problematic and was replaced by a 950-litre (209-gallon) tank. The new variant could be armed with two medium-range R-27T (AA-10 Alamo) air-to-air missiles with infrared guidance or up to six medium-range R-77 (AA-12 Adder) with active radar guidance. The MiG-29SMT also added a range of precision air-to-ground options: Kh-25ML (AS-10 Karen), Kh-29T/L (AS-14 Kedge), Kh-31A/P (AS-17 Krypton) or KAB-500 laser-guided bombs.

The *izdeliye* 9.18 version of the MiG-29SMT with the same N010M radar and other avionic improvements but no spine fuel tank was sold to Yemen, but an upgrade of older airframes to this standard has also been undertaken. This was followed by the *izdeliye* 9.19 version of the MiG-29SMT, which was similar but with the 950-litre (209-gallon) tank in the spine.

In 2006, Algeria ordered 28 aircraft, plus six trainers, but rejected them two years later claiming that they were refurbished rather than new-build airframes. Returned to Russia from 2009, the ex-Algerian aircraft still serve there, designated 9.19R with some upgrades for local use.

There were two prototypes of a two-seat version of the original *izdeliye* 9.17 upgrade, the MiG-29UBT (*izdeliye* 9.52) with a large fuel tank in the spine and revised radome intended to accommodate a small radar. Later, a two-seat MiG-29SMT was produced as the *izdeliye* 9.53, otherwise known as the MiG-29UB modernized (*izdeliye* 9.53) and delivered to Yemen and Algeria. Avionics were similar to the single-seat 9.18/9.19 with a refuelling probe, but with no radar or fuel tank in the spine. Once Algerian *izdeliye* 9.53 aircraft were returned to Russia, they entered service as the *izdeliye* 9.53R.

Indian service

MiG-29UPG (*izdeliye* 9.20) is an advanced mid-life upgrade ordered by India in 2008 for 56 MiG-29 fighters and six MiG-29UB trainers featuring an extended service life and new RD-33 series 3M engines. Six were upgraded by MiG in Russia and the remainder in India. New features include the Zhuk-M2E radar with additional modes, new IRST, a new electronic warfare suite including Indian radar warning receiver and countermeasures dispensers, plus an Italian-made active electronic jammer and Israeli-made jamming pod. The MiG-29UPG incorporates the same spine fuel tank as the 9.19, plus refuelling probe. The two-seat version for India is the MiG-29UPG UB (*izdeliye* 9.53I).

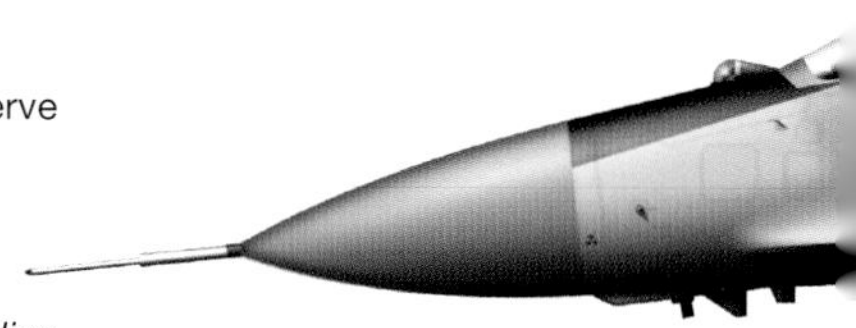

MiG-29 Fulcrum

Weight (maximum take-off): 19,700kg (43,431lb)
Dimensions: Length 16.28m (53ft 5in), Wingspan 11.41m (37ft 5in), Height 4.73m (15ft 6in)
Powerplant: Two Klimov/St Petersburg RD-33 (*izdeliye* 88) turbofans each rated at 49.42kN (11,100lbf) dry, and 81.4kN (18,300lbf) with afterburning
Maximum speed: 2400km/h (1491mph)
Range: 2100km (1305 miles) with under-fuselage drop tank
Ceiling: 18,000m (59,050ft)
Crew: 1
Armament: One 30mm (1.2in) Gryazev-Shipunov GSh-30-1 autocannon in the port leading-edge root extension; up to 4000kg (8,800lb) of weapons and stores

Russian service today

Currently, the Russian Aerospace Force has around 150 single-seaters and 30 or more two-seaters.

Outside Russia, the 'first generation' MiG-29 remains in worldwide service, including with Algeria, Azerbaijan, Bangladesh, Belarus, Bulgaria, Chad, Cuba, Eritrea, India, Iran, Kazakhstan, Libya, Myanmar, North Korea, Peru, Poland, Serbia, Slovakia, Sudan, Syria, Turkmenistan, Ukraine, Uzbekistan and Yemen. Some of these have been upgraded with modernization packages adding advanced multirole capabilities.

MiG-29

'07 Blue' is one of over 200 MiG-29s still in the Russian inventory. This aircraft, normally based at Astrakhan, was one of five MiG-29s performing at Zhukovsky in 2012 during celebrations to mark the 100th anniversary of the Russian Air Force in 2012.

A MiG-29M2 air-superiority fighter of the Russian Aircraft Corporation (commonly known as Mikoyan) takes off near Moscow, 2015.

Mikoyan MiG-29K NATO Reporting name: Fulcrum

A medium-weight multirole carrier fighter, the MiG-29K is superficially similar to the original MiG-29 but uses an all-new airframe.

The project to develop a carrier-capable Fulcrum dates back to the 1980s, with plans to equip an air wing for the Soviet Navy's planned carrier fleet. Compared to the MiG-29, the MiG-29K possesses a larger wing with double-slotted flaps, leading-edge vortex controllers, arrester hook and other naval features.

The carrier-based MiG-29K (*izdeliye* 9.31) was developed alongside the land-based MiG-29M (*izdeliye* 9.15) in the mid-1980s. Both these aircraft featured a new airframe and were intended to have a new N010 radar, now known as the Zhuk. The first of two MiG-29K prototypes took to the air on 23 June 1988 before making its first carrier landing on the then *Tbilisi* on 1 November 1989.

In 1992, after the collapse of the Soviet Union, work on the project was dropped, but when the Indian Navy acquired the former Russian Navy carrier *Admiral Gorshkov*, renamed *Vikramaditya*, work on the MiG-29K resumed. Appearing in late 1999, the new MiG-29K was the *izdeliye* 9.41. Compared to the 9.31, the new version sports increased-area flaps, a new digital fly-by-wire control system and upgraded engines and systems.

The two-seat version is the MiG-29KUB or *izdeliye* 9.47. Both single-seater and two-seater aircraft have the same cockpit canopy with an additional fuel tank in place of the rear seat and crew member in the single-seat version. The first flight of the MiG-29KUB took place on 20 January 2007, followed on 25 June 2007 by the first MiG-29K.

Black Panthers

The first fighters for India were flown in March 2008, and in September 2009 an Indian MiG-29K landed for the first time on the deck of the Russian Navy aircraft carrier *Admiral Kuznetsov* in the Barents Sea. Delivery of the first four MiG-29K fighters to India began in December 2009. These entered service with Indian Naval Air Squadron

MiG-29K Fulcrum

Weight (maximum take-off): 22,400kg (49,384lb)

Dimensions: Length 17.3m (56ft 9in), Wingspan 11.99m (39ft 4in), Height 4.4m (14ft 5in)

Powerplant: Two Klimov RD-33MK afterburning turbofan engines, each rated at 52.96kN (11,905lbf) thrust dry, and 88.3kN (19,840lbf) with afterburning

Maximum speed: 2200km/h (1370mph)

Range: 1850km (1150 miles)

Ceiling: 17500m (57,415ft)

Crew: 1

Armament: One 30mm (1.2in) Gryazev-Shipunov GSh-30-1 autocannon in the port leading-edge root extension; up to 4500kg (8800lb) of weapons and stores

MiG-29K

'34 Blue' is a MiG-29K of the 100th Independent Shipborne Fighter Aviation Regiment based at Severomorsk-3 Air Base near Murmansk in the north of Russia for operations with the Northern Fleet.

(INAS) 303 'Black Panthers' at Indian Naval Station (INS) Hansa in Goa. In March 2010, India ordered another 29 MiG-29K aircraft. The first four from the second contract arrived in India in December 2012.

Russian navy service

In February of the same year, Russia ordered 20 MiG-29KR (*izdeliye* 9.41R) and four MiG-29KUBR (*izdeliye* 9.47R) fighters to serve aboard the sole Russian Navy aircraft carrier, *Admiral Kuznetsov*, the first example of which flew on October 2013. These aircraft are based at Severomorsk alongside Su-33s.

The MiG-29K can carry 4500kg (9921lb) of weapons and stores on nine hardpoints (eight under the wing and one under the fuselage). Air-to-air weapons comprise up to six medium-range R-77 (AA-12 Adder) air-to-air missiles with active radar guidance, or up six short-range R-73 (AA-11 Archer) air-to-air missiles with infrared guidance. Air-to-surface options include four subsonic Kh-35 (AS-20 Kayak) or supersonic Kh-31A (AS-17 Krypton) anti-ship missiles, or four KAB-500Kr TV-guided bombs.

Like its land-based MiG-29 predecessor, the aircraft is fitted with a GSh-30-1 single-barrel 30mm (1.2in) cannon in the port leading-edge root extension.

MiG-29K

Another aircraft of the 100th Independent Shipborne Fighter Aviation Regiment is '42 Blue', shown here with external tanks. Note the long canopy used for both the twin and single seat versions.

Mikoyan MiG-29M and MiG-35

Superficially similar to the original MiG-29, the MiG-29M and MiG-35 feature an all-new airframe to enhance the multirole capability of this medium-weight fighter.

Designed by the Mikoyan Design Bureau in Moscow, both single-seat and two-seat versions use a common airframe and are built by MiG's Lukhovitsy factory, in cooperation with MiG's factory in Moscow and the Sokol factory in Nizhny Novgorod. The MiG-29M and MiG-35 are essentially the same aircraft with different designations depending on marketing.

Development of an advanced multirole version of the MiG-29 originally began in the 1980s, with work on the land-based MiG-29M (*izdeliye* 9.15) taking place in parallel with the carrier-based MiG-29K (*izdeliye* 9.31), with common air-frames and avionics.

Six MiG-29Ms were completed during this period, with the first flying on 25 April 1986. However, the project was abandoned in 1992 after the collapse of the Soviet Union. The MiG-29M was then revived, once again in parallel with the MiG-29K, with changes to the land-based fighter including deletion of the arrester hook and refuelling probe, no wing folding and removal of certain navigation equipment. Up to 6500kg (14,330lb) of weapons and stores can be carried on nine hardpoints (eight under the wing and one under the fuselage).

Air-to-air weapons comprise up to six medium-range R-77 (AA-12 Adder) air-to-air missiles with active radar guidance, or up to six short-range R-73 (AA-11 Archer) air-to-air missiles with infrared guidance. Air-to-surface options include four subsonic Kh-35 (AS-20 Kayak) or supersonic Kh-31A (AS-17 Krypton) anti-ship missiles, four Kh-31P (AS-17 Krypton) anti-radiation missiles, four Kh-38M, four Kh-59M (AS-18 Kazoo) air-to-ground missiles or four KAB-500Kr TV-guided bombs. The MiG-29M retained the GSh-30-1 single-barrel 30mm (1.2in) cannon in the port leading-edge root extension of its MiG-29 forebear.

The two-seat version is the MiG-29M2 (*izdeliye* 9.47S), the first of which flew on 24 December 2011, and was followed by the initial single-seat MiG-29M (*izdeliye* 9.41S) that took to the air on 3 February 2012. This export variant has no arrester hook and other carrier equipment was deleted. First flown on 3 February 2012, the MiG-29M utilizes the N041 (or Zhuk-M1SE for export) radar, OLS-UM infrared search and track (IRST) sensor and

MiG-29M Fulcrum
Weight (maximum take-off): 24,500kg (54,013lb)
Dimensions: Length 17.3m (56ft 9in), Wingspan 11.99m (39ft 4in), Height 4.4m (14ft 5in)
Powerplant: Two Klimov RD-33MK afterburning turbofan engines, each rated at 52.96kN (11,905lbf) thrust dry, and 88.3kN (19,840lbf) with afterburning
Maximum speed: 2100km/h (1300mph)
Range: 2000km (1240 miles)
Ceiling: 16,000m (52,493ft)
Crew: 1 or 2
Armament: One 30mm (1.2in) Gryazev-Shipunov GSh-30-1 autocannon in the port leading-edge root extension; up to 5500kg (12,125lb) of weapons and stores

MiG-35

A pre-production MiG-35 single seater '702 Blue' was constructed in October 2016 and is depicted as it appeared during tests with Russian Aerospace Forces in January 2017.

LEFT AND BELOW:

MiG-29M2

'747 Blue' was one of two prototypes intended for Syria, though the aircraft never went into production due to the deterioration of the political situation in that country. '747 Blue' was retained by RAC MiG and demonstrated at several airshows during 2012 and 2013.

an optional T-220 targeting pod. The MiG-35S (*izdeliye* 9.41SR) is the Russian Aerospace Forces version, similar to the MiG-29M, with N041 radar and uprated RD-33MKR turbofans with the MiG-35UB (*izdeliye* 9.47SR) being the equivalent two-seat derivative for the Russian Aerospace Forces.

Strizhi display team

The first order for the MiG-29M/M2 was reportedly placed by Syria in November 2006 for 12 aircraft (single-seat *izdeliye* 9.61S and two-seat 9.67S), with options for more, but the deal was cancelled in 2015 due to the Syrian civil war. Egypt received 46 MiG-29M/M2 fighters from an order that was placed in 2015, and recently Algeria became a customer with reports from late 2020 that the first MiG-29M/MiG-29M2 aircraft had been delivered. Reportedly, a total of 14 aircraft have been acquired. Russia outlined plans to purchase 'up to 37' MiG-35S fighters, broadly similar to the MiG-29M. However, so far firm orders have only been placed for six examples, four single-seaters and a pair of two-seaters for evaluation. Thereafter, they are expected to join the *Strizhi* (Swifts) display team.

MiG-35

The MiG-35 (*izdeliye* 9.61) was the version unsuccessfully offered to India for the Medium Multi-Role Combat

MiG-29M

Much photographed due to its appearance in the West at the 1992 Farnborough Airshow, '156 Blue' was one of the first MiG-29Ms constructed before the fall of the Soviet Union which led to the abandonment of the MiG-29M programme for nearly 20 years.

Aircraft (MMRCA) competition, which had called for 126 fighters. It is broadly similar to MiG-29M but with a Zhuk-AE radar with active electronically scanned array (AESA). Only one was built, based on a converted MiG-29K. Together with the two seat MiG-35D (*izdeliye* 9.67) converted from a MiG-29KUB airframe and also fitted with Zhuk-AE radar, this aircraft was evaluated in India during 2009. Ultimately, the MiG-35 was converted back to a MiG-29K and the MiG-35d returned to MiG-29KUB configuration.

Export model

A new 'renewed export configuration' MiG-35 was unveiled at the MAKS International Aviation and Space Salon in Moscow in 2019. This version reinstates the Zhuk-AE AESA radar and has increased-area wing and tail surfaces. Again, it is likely aimed at India's ongoing fighter requirements.

Mikoyan MiG-31 NATO reporting name: Foxhound

One of the world's fastest combat aircraft, the MiG-31 was designed in the Soviet era by the Mikoyan Design Bureau in Moscow as a long-range Mach 2.8 interceptor.

Initially considered a relatively straightforward two-seat MiG-25 variant featuring multi-target capability and extended endurance designated as the MiG-25MP, the programme to develop what would become the extensively redesigned MiG-31 was launched in May 1968. The first flight of the MiG-25MP occurred on 16 September 1975 and the initial production MiG-31 appeared in late 1976, with series production undertaken by the Sokol factory in Nizhny Novgorod (formerly Gorky).

First production version

The MiG-31 was officially commissioned into service with the Soviet Air Defence Forces in May 1981 with initial operational capability, first achieved by the MiG-31 unit at Pravdinsk, being attained during 1983. In its initial production version the MiG-31 (*izdeliye* 01) featured the Zaslon fire-control system, the world's first operational electronically scanned phased-array fighter radar possessing a maximum detection range of 200km (124 miles) with a typical width of detection of 225km (140 miles). Initially, the MiG-31 carried four semi-active

MiG-31M

The first MiG-31 seen in the West, '057 Blue' was the seventh MiG-31M to be built and included the improved Zaslon-M radar system, six K-37, four K-77 AAMs, and a modified airframe with around two tons of additional fuel and more powerful D-30D-6M engines. Ultimately the ambitious 'Foxhound-B' fell victim to the huge funding cuts in the wake of the collapse of the Soviet Union.

MiG-31M Foxhound

Weight (maximum take-off): 46,200kg (101,853lb)
Dimensions: Length 20.62m (67ft 8in) without probe, Wingspan 13.46m (44ft 2in), Height 6.45m (21ft 2in)
Powerplant: Two Soloviev D-30F6 afterburning turbofan engines, 93kN (21,000lbf) thrust each
Maximum speed: Mach 2.35
Range: 3300km (2051 miles) unrefuelled
Ceiling: 20,600m (67,585ft)
Crew: 2
Armament: One six-barrel GSh-6-23 23mm (0.9in) cannon, four semi-active radar-guided R-33 AAMs below the fuselage, plus two medium-range R-40TD or four R-60M AAMs on wing pylons

MiG-31M

Viewed from the front, the MiG-31 airframe is dominated by the truly enormous intakes for its two Aviadvigatel/Perm D-30F-6 turbofans.

radar-guided R-33 (AA-9 Amos) or R-33S air-to-air missiles carried in tandem pairs under the fuselage, plus two medium-range infrared-guided R-40TD (AA-6 Acrid) or four infrared-guided R-60M (AA-8 Aphid) self-defence air-to-air missiles on the wing pylons. The aircraft featured a six-barrel 23mm (0.9in) GSh-6-23 cannon in the starboard fuselage with 260 rounds.

As designed, four MiG-31s were intended to be able to defend an area of airspace along a front of around 800-900km (497-559 miles).

Interceptor role

The MiG-31 replaced the MiG-25 and Tu-128 in the long-range interceptor role and boasted almost double the range of the MiG-25 and far more capable sensors and weapons than the Tu-128. It was designed to protect the northern and far eastern borders of the USSR against aerial incursion and was specifically intended to defend against cruise missile attack as part of a ground-controlled intercept (GCI) network, although it is also able to operate autonomously. The advanced MiG-31M Foxhound-B interceptor and the anti-satellite MiG-31D were developed in the 1980s, but were cancelled in the early 1990s. Later modifications of the MiG-31 have been developed for long-range strike and anti-satellite missions.

Total production between 1976 and 1994 amounted to 519, of which 349 were of the initial MiG-31 version followed by 101 MiG-31DZ (*izdeliye* 01DZ) that introduced an inflight refuelling probe and entered production in 1989. These were followed by 69 MiG-31B versions featuring improved Zaslon-A radar and nuclear-tipped

MiG-31BM

Based at Savasleyka air base, a long-term Foxhound operating location, '82 Blue' serves with the 4th Guards Kerch Red Banner Aviation Squadron (4 GvTsPAP). This unit is the crew training and combat application unit for the MiG-31.

MiG-31B/BS Foxhound
Weight (maximum take-off): 46,200kg (101,853lb)
Dimensions: Length 20.62m (67ft 8in) without probe, Wingspan 13.46m (44ft 2in), Height 6.15m (20ft 2in)
Powerplant: Two Aviadvigatel/Perm D-30F-6 turbofans, each rated at 152kN (34,172lb) thrust with afterburning
Maximum speed: Mach 2.83
Range: 3300km (2051 miles) unrefuelled
Ceiling: 20,600m (67,585ft)
Crew: 2
Armament: One six-barrel GSh-6-23 23mm (0.9in) cannon, four semi-active radar-guided R-33 AAMs below the fuselage, plus two medium-range R-40TD or four R-60M AAMs on wing pylons

R-33S missiles, while some earlier aircraft were upgraded to the same standard and designated the MiG-31BS (*izdeliye* 01BS). Almost all MiG-31Bs have now received a mid-upgrade to MiG-31BM standard (*izdeliye* 01BM), developed from early 2000s with Zaslon-AM fire-control system with upgraded radar, new LCD cockpit displays, communication and navigation equipment.

The MiG-31BM replaced the R-33s with four radar-guided R-37M (AA-13 Axehead) air-to-air missiles with increased range, and the R-40s and R-60s were replaced with up to four short-range infrared-guided R-73 (AA-11 Archer) AAMs. The first upgraded MiG-31BM in current configuration took to the air in September 2005. By 2014, around 50 aircraft had been upgraded and work continues.

Current Russian service

Today, Russian Aerospace Forces operate 130 MiG-31s, plus another squadron assigned to the Russian Navy's Northern Fleet, and the aircraft is expected to remain in service until at least 2030. A further 130 or so aircraft are in storage facilities, some gradually being returned to service after refurbishment as MiG-31BM versions. The only foreign operator is Kazakhstan, which currently operates between 20 and 30 ex-Soviet aircraft at Karaganda.

The MiG-31K (for *Kinzhal*, or dagger) is the long-range strike version, produced via modification of earlier interceptor versions that was reportedly in service at Akhtubinsk by early 2018. Its role is to launch a single *izdeliye* 292 (Western reporting name Killjoy) ballistic missile from the centreline. This huge weapon is developed from the 9M723 Iskander (SS-26 Stone) surface-to-surface missile. Flying at hypersonic speed and manoeuvring during its flightpath, the missile is intended to defeat anti-missile defence systems and deliver nuclear and conventional warheads over (reported) distances of more than 2000km (1243 miles).

Anti-satellite system

Meanwhile, tests of the Burevestnik (storm petrel) anti-satellite system began in September 2018 as the successor to the Cold War-era MiG-31D programme, using a MiG-31 to carry a single *izdeliye* 293 missile again launched from the centreline. Burevestnik is an airborne anti-satellite (ASAT) system that can conduct space strikes to disrupt enemy military control, communications or information support. The two-stage solid-fuel Burevestnik is believed to deliver small interceptor satellites into orbit, which in turn are used to combat enemy satellites.

Sukhoi Su-24 NATO reporting name: Fencer

The Su-24 was developed during the Cold War as a nuclear and conventional tactical bomber, with missions including air interdiction and defence suppression.

Work on what became the Su-24 began in 1963 to provide a strike aircraft broadly analogous to the British TSR.2 and American F-111. As well as high speed and low-altitude performance, it was originally planned to provide short take-off and landing (STOL) capabilities and the T-58M (later designated T-6) therefore combined a delta wing with four lift engines in the fuselage. In this form the T6-1 prototype was first flown on 2 July 1967. The lift jets used too much internal space and were discarded from the revised second prototype T6-2I, which also replaced the delta with a variable-geometry wing and first flew on 17 January 1970.

By December 1971 the first production Su-24 Fencer-A had been completed at the Novosibirsk Aircraft Plant and the aircraft was officially commissioned into service in February 1975. The first 15 Su-24 Fencer-As featured a box-like rectangular rear fuselage concealing the engine nozzles. These were followed by the Fencer-B, with enlarged drag-chute container and a reprofiled, rounded rear fuselage and Fencer-C with a new radar warning receiver, long-range navigation system and identification friend or foe (IFF). These two models comprised the first 27 production batches, although these have all now been retired from Russian service. The last first-generation Su-24s were

Su-24M2

Operated by the 277th Bomber Aviation Regiment, RF-95108 77 is one of a relatively small number of 'Fencers' upgraded to Su-24M2 standard. The aircraft is shown launching an example of the Kh-25ML (AS-10 'Karen') laser-guided air-to-surface missile.

completed in 1983, by which time the definitive Su-24M (Fencer-D) was coming off the production line.

The Su-24M originated in a 1975 requirement for a much-improved second-generation bomber, or T-6M, with a wider array of precision-guided weapons. This required a new fire-control system and the aircraft also added a more comprehensive self-defence system and inflight refuelling capability. The first production Su-24M appeared in 1979 and could be distinguished from the basic Su-24 by its nose, which was lengthened by 750mm (29in). The Su-24M carries up to 7500kg (1653lb) of stores on seven external pylons (three under the fuselage, two under the wing gloves and two pivoting pylons under the outer wing panels).

Radiation detection

When the Su-24M entered service during the Cold War, its main weapons for the strike role were tactical nuclear bombs. Conventional options include various precision-guided weapons, freefall bombs, sub-munitions dispensers, incendiary tanks, rockets and gun pods. Two infrared-guided R-60M (AA-8 Aphid) air-to-air missiles can be carried for self-defence and the aircraft features a six-barrel 23mm

Su-24M2

Weight (maximum take-off): 39,700kg (87,523lb)

Dimensions: Length 24.53m (80ft 6in) with probe, Wingspan 10.37m (34ft) fully swept, Height 6.19m (20ft 3in)

Powerplant: Two Lyulka-Saturn AL-21F3 turbojets, each rated at 108.36kN (24,361lb) thrust with afterburning

Maximum speed: Mach 1.35

Range: 4270km (2653 miles), with one inflight refuelling

Ceiling: 11,500m (37,730ft)

Crew: 2

Armament: One 23mm (0.9in) six-barrel GSh-6-23M cannon, plus a maximum external load of 7500kg (16,535lb) carried on seven external pylons, including tactical nuclear and conventional bombs, precision-guided munitions, submunitions dispensers, rockets and gun pods

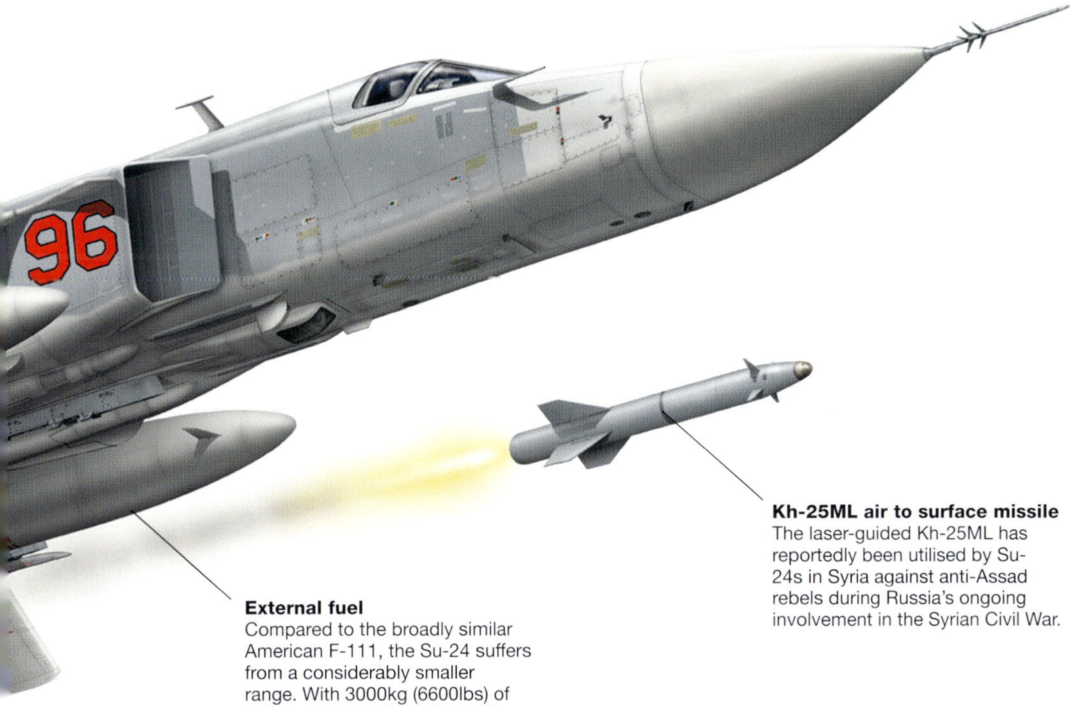

Kh-25ML air to surface missile
The laser-guided Kh-25ML has reportedly been utilised by Su-24s in Syria against anti-Assad rebels during Russia's ongoing involvement in the Syrian Civil War.

External fuel
Compared to the broadly similar American F-111, the Su-24 suffers from a considerably smaller range. With 3000kg (6600lbs) of ordnance and external tanks the combat radius is 615km (382 miles) at low level.

(0.9in) GSh-6-23M cannon in the starboard lower fuselage with 500 rounds. The Su-24M served as the basis for the all-weather, day and night, multi-sensor tactical reconnaissance Su-24MR, first flown in May 1980. The basis of the reconnaissance version is the BKR-1 reconnaissance complex comprising a synthetic aperture side-looking radar in the aircraft nose, a TV camera that transmits pictures to ground stations in real time, infrared line scanner, vertical daylight camera and oblique daylight camera. Laser reconnaissance, ELINT and radiation detection capabilities can all be provided by podded sensors.

Production totalled between 1300 and 1400 Su-24s during 1972–92. This total includes around 500 Su-24Ms, nearly 200 reconnaissance Su-24MRs and more than 100 Su-24MK export versions with no nuclear capability and minor differences in radio and IFF. Delivery of the first Su-24 bombers took place during 1973 to the crew conversion and evaluation centre at Lipetsk. The first operational unit was the 4th Bomber Aviation Regiment at Chernyakhovsk on the Baltic Sea, which was declared service ready in May 1974.

First combat use occurred in Afghanistan when Su-24s were used for high-level bombing in 1984. Su-24 and Su-24M bombers, as well as Su-24MR reconnaissance aircraft, were used by Russia over Chechnya from 1994 employing freefall bombs and unguided rockets, as well as KAB-500Kr guided bombs and guided missiles. Lessons from the Chechen campaign led to the SVP-24 upgrade developed by the Gefest & T company featuring improved accuracy of the navigation system supposedly giving freefall bombs the accuracy of smart munitions. Meanwhile, Sukhoi's own Su-24M2 (T-6M2) mid-life upgrade resulted in series deliveries from 2007 but was reportedly much more expensive than the SVP-24.

Bomber versions

Currently, Russian Aerospace Forces have around 150 bomber versions (including around 50 SVP-24 and 30 M2 upgraded variants) plus 80 reconnaissance versions. The Russian Navy has around 28 more bombers plus eight reconnaissance versions, although the aircraft is gradually being replaced by the Su-34 and Su-30SM. Many more Su-24s were left in the hands of former Soviet states, of which Ukraine is now the final operator.

The first export Su-24MK bombers were delivered to Iraq in 1988, many of these later fleeing to Iran, which put them into service. In the Soviet era, other Su-24MKs were sold to Algeria, Libya and Syria. More recently, Angola has received at least one Fencer, while ex-Belarusian examples were sold to Sudan. Since September 2015, Russian Su-24Ms have been stationed in Syria to support the regime in the ongoing civil war. One Su-24M was shot down by a Turkish F-16C after reportedly straying into Turkish airspace in November 2015.

On approach to Dyagilevo air base, Razan, Russia, in 2016, this well-used Su-24M was armed only with a pair of unguided rocket pods. The sturdy undercarriage and prominent mudguard on the nose gear are obvious in this view.

Sukhoi Su-25 NATO reporting name: Frogfoot

The Su-25 single-seat subsonic close air support aircraft features heavy armour protection and a good degree of manoeuvrability, conferring upon it the ability to deliver accurate attacks against small ground targets.

A pair of Su-25UB two-seaters take off from Kubinka air base, near Moscow. The object above the rear of the canopy is a periscope allowing the occupant of the rear seat to see ahead.

Developed by the Sukhoi Design Bureau in Moscow, work on what became the Su-25 began in 1968 and the prototype T8-1 first flew on 22 February 1975. Two test aircraft, the T8-1D and T8-3, were deployed to Afghanistan for trials in 1980 before 12 aircraft were delivered to the first operational unit, the 200th Independent Attack Aviation Squadron (OShAE) at Sital Chay air base, in Azerbaijan in April 1981. The Su-25 saw the World War II-era *Shturmovik* classification reintroduced to the Soviet Air Force, these units last having flown the Il-10 in the 1950s.

Frogfoot-A

Initial production was the Su-25 Frogfoot-A, the standard single-seater that was offered for export as the Su-25K. This was followed by the Su-25BM that added provision for target towing and was fitted with uprated R-195 engines. The export version of this variant was designated Su-25BMK. The Su-25T and Su-25TM, dedicated anti-armour versions with a revised attack/navigation system, electro-optical sight, provision for Vikhr anti-tank guided missiles and other guided weapons, were first flown on 17 August 1984.

In total, 20 Su-25Ts were manufactured in Tbilisi in the early 1990s, followed by two Su-25TM at Ulan-Ude, but were mainly used for evaluation. Su-25U is a Su-25T converted into a two-seat trainer version by the Tbilisi factory, of which two were acquired by the Georgian Air Force.

Combat versions

Single-seat combat versions of the Su-25 were manufactured at the Tbilisi aircraft plant between 1979 and 1992, while two-seaters came from the Ulan-Ude (Russia) factory between 1987 and 1992. Initial two-seaters were the Su-25UB Frogfoot-B, tandem two-seat combat trainer in which the weapons systems were retained and which featured an enlarged vertical fin compared to the single seaters.

The export equivalent of the initial production two-seater was designated the Su-25UBK. The excellent handling characteristics of the Su-25 have seen the two-seater utilized by the Russian Na-vy as a training machine, the Su-25UTG. This is a carrier-capable trainer with weapons system removed and an instrument landing system and arrester hook added. The small batch completed was operated aboard the aircraft carrier *Admiral Kuznetsov,* with survivors used by the Navy for land-based training.

Su-25SM3

'93 Red' (formerly '05 White' before its upgrade to Su-25SM3 standard) is shown as it appeared during 2021. The short overall length and comparatively large vertical tail surface are evident, as is the muzzle of the Frogfoot's primary weapon, the 30mm GSh-2-30 cannon in the lower nose.

Both single- and two-seaters are equipped with eight main wing hardpoints, each able to carry a 500kg (1102lb) load, while two small outboard pylons are available for R-60 (AA-8 Aphid) air-to-air missiles only. Maximum weapons load is 4340kg (9568lb). The most advanced weapons comprise laser-guided Kh-25ML (AS-10 Karen) and Kh-29L (AS-14 Kedge) air-to-surface missiles but unguided rockets, freefall bombs, sub-munitions dispensers, incendiary tanks and gun pods can all be carried. Russian Air Force Su-25s are nuclear-capable. Su-25s are armed with one GSh-2-30 twin-barrel 30mm (1.2in) cannon in the port side of the front fuselage below the cockpit, with 250 rounds.

Pictured on approach to land at Kubinka air base after participating in the 2020 Victory Day parade, Su-25BM '74 Red' has deployed its landing lights from its distinctive flattened wingtip pods, the rear part of which contain airbrakes.

A mid-life upgrade programme was launched in 1999 based around new PrNK-25SM mission avionics, including Kopyo radar and modernized cockpit with two large digital displays. Eventually, Russia opted for a more modest option, without radar, also known as Su-25SM. The initial upgraded version is Su-25SM1, followed by Su-25SM2 and, from 2014, the definitive Su-25SM3. All Russian single-seat aircraft are planned to be upgraded to Su-25SM standard. The upgraded Su-25SM can carry up to 5000kg (11,023lb) on the eight main hardpoints. New weapons include TV-guided KAB-500Kr bombs, S-25 laser-guided rockets and R-73 (AA-11 Archer) air-to-air missiles.

About 1320 Su-25s were built, but the Ulan-Ude plant retains the infrastructure to build more Su-25s if ordered. Russia currently has some

Su-25SM

Weight (maximum take-off): 19,000kg (41,888lb)
Dimensions: Length 15.53m (50ft 11.5in), Wingspan 14.36m (47ft 1in), Height 4.80m (15ft 9in)
Powerplant: Two non-afterburning Soyuz/Moscow R-95Sh turbojets each developing 40.21kN (9,039lb) of thrust
Maximum speed: Mach 0.82
Range: 1850km (1450 miles) ferry range with drop tanks
Ceiling: 7000m (22,966ft)
Crew: 1
Armament: One 30mm (1.2in) twin-barrel GSh-2-30 cannon, plus a maximum external load of 4340kg (9568lb) carried on eight main underwing hardpoints and two smaller pylons. Precision ordnance includes laser-guided Kh-25ML (AS-10 'Karen') and Kh-29L (AS-14 'Kedge') air-to-surface missiles and TV-guided KAB-500Kr bombs

200 Su-25s, including 20 Su-25UB trainers in service. Other post-Soviet operators are Armenia, Azerbaijan, Belarus, Georgia, Kazakhstan, Turkmenistan, Ukraine and Uzbekistan, all of which inherited the aircraft from the Soviet Union. About 210 aircraft were exported outside the former USSR; some of these still serve in Angola, Bulgaria, Chad, Democratic Republic of the Congo, Equatorial Guinea, Eritrea, Ethiopia, Gambia, Iran, Iraq, North Korea, Peru and Sudan. Former operators that have retired the type include the Czech Republic, Ivory Coast, Macedonia and Slovakia.

African campaigns

From June 1981, Su-25s with the Soviet 200th Independent Attack Aviation Squadron were deployed to Shindand, Afghanistan, to undertake combat trials. Subsequently, Russian Frogfoots saw combat in the first and second Chechen wars in Georgia in 2008, and, since September 2015, in Syria. Ukrainian Su-25s saw action in the campaign in the east of that country during 2014 before heavy losses saw their withdrawal from that theatre. The aircraft has been used by both sides in Russia's invasion of Ukraine, with losses suffered by aircraft of both nations. The type has also been widely used in combat by export operators, including Iraq, first against Iran in the 1980s and most recently in its campaign against the so-called Islamic State. The Frogfoot has taken part in numerous campaigns in Africa, including operations by Angola, Chad, Ivory Coast and Sudan. More recently, both Armenian and Azeri Su-25s saw action during the 2020 conflict in Nagorno-Karabakh.

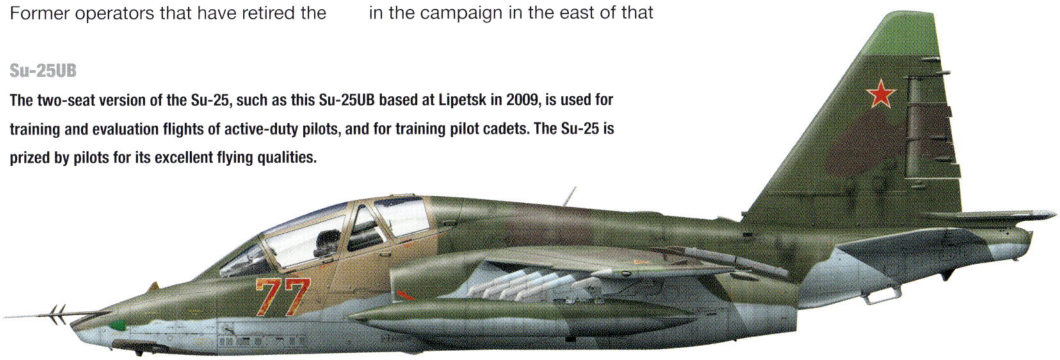

Su-25UB

The two-seat version of the Su-25, such as this Su-25UB based at Lipetsk in 2009, is used for training and evaluation flights of active-duty pilots, and for training pilot cadets. The Su-25 is prized by pilots for its excellent flying qualities.

Sukhoi Su-27 NATO reporting name: Flanker

Designed as a response to the American F-15, the Su-27 was the most potent fighter developed by the Soviet Union and remains in widespread service today.

For many years the Su-27 was the most important fighter in the Russian inventory, although its importance is now diminishing. A 'heavy' long-range air-superiority fighter and escort fighter, it initially fulfilled only a secondary air-to-ground role with exclusively unguided ordnance. However, after the Su-27SM upgrade the aircraft became a true multirole fighter.Beginning life as the Perspektivnyi Frontovoi Istrebitel, or Future Tactical Fighter programme, it was launched in 1969 in response to the McDonnell Douglas F-15, and Sukhoi's T-10 proposal was selected in 1971.

Maiden flight

The first prototype T-10-1 made its maiden flight on 20 May 1977. After seven T-10s (Flanker-A) were completed the design was thoroughly revised, leading to the T-10S. The first prototype T-10S-1 made its maiden flight on 20 April 1981, with production aircraft following from June 1982 in two basic forms, the initial Su-27 that was quickly supplanted by the Su-27S with improved AL-31F engine and the Su-27P air defence version, which dispensed with the air-to-ground capability of the basic aircraft, both of which were known as Flanker-B to NATO.

The Su-27UB combat trainer (known as Flanker-C to NATO) with the same fire-control system and weapons of the single-seat version but reduced fuel capacity appeared in 1986, and in the same year a unique stripped-down version designated the P-42 appeared that successfully beat several time-to-height records set by the F-15 'Streak Eagle'.

A total of 645 single-seat Su-27s was completed by 1999, plus 190 Su-27UB/UBK two-seaters. Production after this was date limited, as new Flanker versions were now prioritized. Nonetheless, after 1999 five aircraft were completed for Indonesia and 12 Su-27SM(3)s were built for Russia.

Current service numbers

Service entry with the Soviet Air Force was at Dzyomgi, where initial operational capability was declared on the type in December 1984. Today, the Russian Aerospace Forces retain around 100 Su-27s complemented by around 20 trainers.

The first export customer for the Su-27 was China, which received 36 Su-27SK single-seaters and 10 Su-27UBK two-seaters in two batches in 1992 and 1996 before purchasing an additional batch of 28 Su-27UBKs in 2000-02. In 1996 China also acquired a licence to build Su-27s under the local designation J-11. After 105 of these were constructed to Su-27 standard, China switched to the fully indigenous J-11B version, followed by other Chinese derivatives.

Su-27P

Weight (maximum take-off): 28,300kg (62,391lb)
Dimensions: Length 21.94m (72ft) without probe, Wingspan 14.7m (48ft 3in), Height 5.93m (19ft 6in)
Powerplant: Two Saturn AL-31F turbofans, each rated at 122.58kN (27,558lb) thrust with afterburning
Maximum speed: Mach 2.35
Range: 3720km (2312 miles)
Ceiling: 18,500m (60,700ft)
Crew: 1
Armament: One GSh-301 30mm (1.2in) cannon, plus up to 4430kg (9766lb) of external stores on 10 weapons pylons

Su-27P

An example of the dedicated air defence 'Flanker', RF-92407 '18 Blue' of the Russian Aerospace Forces is pictured as it appeared in July 2021.

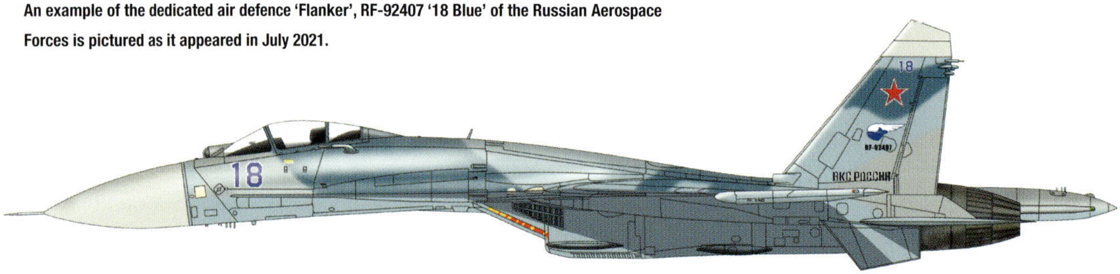

The demise of the Soviet Union left Su-27s in the hands of Belarus (now retired), Ukraine and Uzbekistan. Armenia and Kazakhstan later received second-hand aircraft from Russia. Angola, Ethiopia and Eritrea have also received small numbers of Su-27s from ex-Soviet stocks.

In addition to China, new-build export Su-27s were delivered to Vietnam (seven Su-27SK: an export version with a downgraded fire-control system, and five Su-27UBK, the export two-seater in two batches) and Indonesia (two Su-27SKs and three Su-27SKMs, an upgraded export version with a fire-control system based on that used in the multirole Su-30MK2).

Airframe upgrade

A Russian upgrade for the Su-27 was launched in 2003, bringing the aircraft to Su-27SM standard. In addition to the modernization of existing airframes, a batch of 12 newly built Su-27SM(3) versions was also completed that feature a revised fire-control system and R-77-1 missile capability among other improvements.

The basic Su-27 can carry up to 4430kg (9766lb) of external stores on 10 pylons. Air-to-air weapons include up to six semi-active radar-guided medium-/extended-range R-27R/ER (AA-10 Alamo) air-to-air missiles, two infrared-guided medium-/extended-range R-27T/ET air-to-air missiles, two passive-radar R-27P/EP AAMs and up to six infrared-guided short-range R-73 (AA-11 Archer) air-to-air missiles. Unguided air-to-ground weapons include conventional bombs, rockets and freefall nuclear bombs.

Su-27SM

In the Su-27SM, an upgraded fire-control system allows the use of guided air-to-ground weapons, including Kh-31A/P (AS-17 Krypton) anti-ship/anti-radar missiles, Kh-29L/T/TD (AS-14 Kedge) laser-/TV-guided missiles, S-25LD laser-guided rockets and KAB-500Kr and KAB-1500Kr guided bombs.

New air-to-air options include the R-77 and R-77-1 (AA-12 Adder) medium-range missile with active radar guidance. The Su-27 is also armed with a GSh-30-1 single-barrel 30mm (1.2in) cannon in the starboard leading-edge root extension.

Recent deployments

The Su-27 has seen combat service during the 1992–93 war in Abkhazia against Georgian forces and the 2008 South Ossetia War. More recently, a squadron of Su-27SM3s was deployed to Syria in November 2015 as part of the Russian military intervention.

Sukhoi Su-27SKM

Weight (maximum take-off): 28,300kg (62,391lb)
Dimensions: Length 21.94m (72ft) without probe, Wingspan 14.7m (48ft 3in), Height 5.93m (19ft 6in)
Powerplant: Two Saturn AL-31F turbofans, each rated at 122.58kN (27,558lb) thrust with afterburning
Maximum speed: Mach 2.15
Range: 3680km (2285 miles)
Ceiling: 18,000m (65,055ft)
Crew: 1
Armament: One GSh-301 30mm (1.2in) cannon, plus up to 8000kg (17640lb) of external stores on 10 weapons pylons

Su-27SKM

Formerly an Su-27SK, '305 Blue' was rebuilt in 2002 as demonstrator for Su-27SM and export Su-27SKM was developed for export with new avionics and air to ground missiles. This aircraft was part of the flying display at the 2005 Paris Airshow in this striking three-colour camouflage scheme. Five examples of the Su-27SKM were subsequently obtained by Indonesia, which remains the sole operator of this variant.

Cockpit
Compared to contemporary Western aircraft, the Su-27 featured more analogue instrumentation and presented less information to the pilot via the Head Up Display (HUD). However, pilots could aim and fire missiles via a helmet-mounted sight within a wide hemisphere in front of the aircraft, rather than simply straight ahead, a potentially decisive feature unavailable to US pilots until 2003.

Su-27 'Flanker B'

One of the later early-model Flankers to be built, '24 Blue' was on the strength of the 582nd Guards Fighter Air Regiment, stationed at Chojna air base in Poland in 1991, one of two Polish Frontal Aviation bases utilized by 'Flankers' before the fall of the Soviet Union.

Su-27 'Flanker-B'

Weight (maximum take-off): 28,300kg (62,391lb)
Dimensions: Length 21.94m (72ft) without probe, Wingspan 14.7m (48ft 3in), Height 5.93m (19ft 6in)
Powerplant: Two Saturn AL-31F turbofans, each rated at 122.58kN (27,558lb) thrust with afterburning
Maximum speed: Mach 2.35
Range: 3720km (2312 miles)
Ceiling: 18,500m (60,700ft)
Crew: 1
Armament: One GSh-301 30mm (1.2in) cannon, plus up to 4430kg (9766lb) of external stores on 10 weapons pylons

Radar
The N001 Mech (Sword) radar was developed by the Tikhomirov Scientific Research Institute of Instrument Design specifically for use in the Su-27. Initially disappointing, the radar scraped through its state trials and from 1991 was fitted to the 'Flanker' largely because there was no alternative set available. It was subsequently much improved.

Tailfins
The two large vertical tailfins of the Su-27 were originally mounted atop the engine nacelles on the prototype T-10. By the time the aircraft entered production, they had been moved to their definitive position outboard of the engines.

Missiles
In 1991 the Su-27 utilized the R-27 (known as the AA-10 'Alamo' to NATO) and R-73 (AA-11 Archer) for long and short range use respectively. The R-73 seen here being launched had a 40 degree off boresight capability, later increased to 60 degrees in the R-73M.

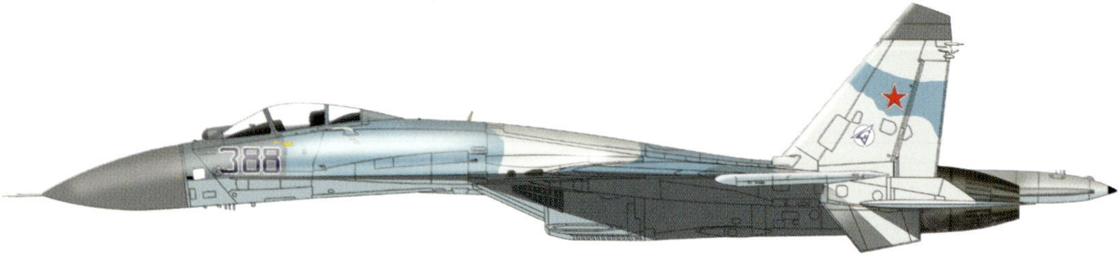

Su-27

'388 Blue' was displayed at the Farnborough Airshow as early as 1990, causing something of a stir with its then unmatched manoeuvrability.

Su-27UB

The Russian Knights display team has operated both single and twin seat 'Flankers' for many years. '19 Blue', an Su-27UB, was another Soviet era Flanker which visited RAF Scampton in the UK in 1991.

Su-27UB 'Flanker-C'

Weight (maximum take-off): 30,500kg (67241lb)
Dimensions: Length 21.94m (72ft) without probe, Wingspan 14.7m (48ft 3in), Height 5.93m (19ft 6in)
Powerplant: Two Saturn AL-31F turbofans, each rated at 122.58kN (27,558lb) thrust with afterburning
Maximum speed: Mach 2
Range: 3000km (1864 miles)
Ceiling: 17,500m (57,415ft)
Crew: 2
Armament: One GSh-301 30mm (1.2in) cannon, plus up to 4430kg (9766lb) of external stores on 10 weapons pylons

A Sukhoi Su-27SM number RF-92209 takes off at Kubinka air force base, Moscow region, June 2015.

Sukhoi Su-30 NATO reporting name: Flanker

The potent, two-seat, multirole long-range Flanker family consists of two distinct lines of development, although both are known as the Su-30.

Developed by the Sukhoi Design Bureau in Moscow, series production of the Su-30 is undertaken at two separate facilities in Irkutsk and Komsomolsk-on-Amur. The Irkutsk Su-30 is built by the Irkutsk aircraft plant, which is part of the Irkut Corporation. This line began with the Su-30MKI for India, developed and built in cooperation with Hindustan Aircraft Limited (HAL).

The Irkutsk line is more advanced: these aircraft are characterized by their canard foreplanes, thrust-vectoring engines and an open-architecture fire-control system based around the electronically scanned Bars radar. By contrast, the 'Komsomolsk Su-30' is manufactured by the Komsomolsk-on-Amur aircraft plant, or KnAAPO, which is part of the Sukhoi company.

The Komsomolsk line began with the Su-30MKK made for China. Su-30s from KnAAPO are based on a standard Su-27UB airframe, albeit strengthened for heavier payloads, and closed-architecture fire-control system, although successive upgrades have been incorporated. Both Su-30 lines are primarily long-range fighters, although in Russian service both the Su-30SM and Su-30M2 have been used for advanced training in addition to their combat roles.

The concept behind the Su-30 began with the two-seat Su-27UB Flanker-C combat trainer, which retained operational capability but possessing the same fire-control system and weapons of the single-seat Su-27. The original Su-30 (also known as the Su-27PU) simply added a tactical situation display in the rear cockpit and an inflight refuelling probe, the prototype first flying on 30 December 1989. Only five Su-30s were completed for the then Russian Air Force in the mid-1990s due to post-Soviet budget restrictions.

Meanwhile, the export-oriented Su-30MK (*Mnogofunktsionalnyi, Kommercheskiy*, multifunction, commercial) added multirole capability to the Su-30 airframe. In 1996, India selected the Su-30MKI (I for India) with canard foreplanes, a digital fly-by-wire system, Israeli avionics and AL-31FP thrust-vectoring engines, the N011M Bars passive electronically scanned radar, helmet-mounted sight and targeting pod, plus Indian-made identification friend or foe (IFF).

The prototype Su-30MK-1 (T-10PMK-1) first flew in July 1997, followed by the first fully equipped Su-30MKI in November 2000.

Export versions

Pending availability of the fully equipped Su-30MKI, Irkutsk provided India with 18 Su-30K aircraft between 1997 and 1999 – export versions of the basic Su-30.

Sukhoi Su-30MKI

Weight (maximum take-off): 34,000kg (74,957lb)
Dimensions: Length 21.94m (71ft 11in), Wingspan 14.7m (48ft 2in), Height 6.4m (20ft 11in)
Powerplant: Two AL-31FP (*izdeliye* 96) thrust-vectoring turbofans, each providing 122.6kN (27,550lbf) thrust at maximum afterburning
Maximum speed: 2120km/h (1320mph)
Range: 3000km (1864 miles)
Ceiling: 17,300m (56,758ft)
Crew: 2
Armament: One 30mm (1.2in) Gryazev-Shipunov GSh-30-1 Gast autocannon; up to up to 8130kg (17,920lb) of weapons and stores

Su-30SM

One of the Russian Aerospace Force's Su-30SM fleet, based at Dyagilevo air base in 2018. Russian Aerospace Forces possess around 140 Su-30s, with Naval Aviation units adding around 50 more.

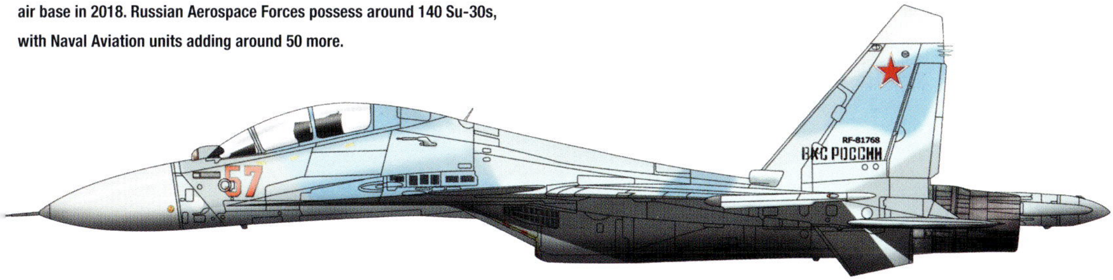

These were replaced by new Su-30MKIs during 2008 and 2009, while the Su-30Ks were sent to Belarus for overhaul, and later delivery to Angola (12 aircraft). Deliveries of Su-30MKIs began in June 2002; successive orders have been placed for 284 examples, of which 50 were delivered from Irkutsk and others built locally by HAL.

The next customer was Malaysia, with 18 Su-30MKM fighters in which modifications were made to the Bars-M radar, infrared search and track (IRST) sensor and helmet-mounted sight. Israeli subsystems found in the Su-30MKI were replaced by French equipment including Thales's Damocles targeting pod. These aircraft were delivered between 2007 to 2009. Algeria ordered 28 Su-30MKI(A) variants based on the Su-30MKI in 2006, before adding another order for 16 in 2010 and 14 more in 2016.

Russian Aerospace Forces

In 2012, Russia ordered 30 Su-30SM aircraft that feature modified Bars-R with Indian-made computers, Russian-made IFF and radios, but other French-made equipment was retained. Another 30 were ordered in 2012, followed by the first batch for the Russian Navy in 2013. The first Su-30SM made its maiden flight on 21 September 2012 and the first deliveries were to Domna air base in November 2013. To date, more than 100 Su-30SMs have been delivered to the Russian Aerospace Forces and Russian Navy, including some for the Russian Knights aerobatic team. The most recent Russian deliveries, as of 2022, comprise improved Su-30SM2 aircraft for the Russian Navy, 21 of which are on order.

The Su-30SM2 is built as new, as well as an upgrade for the Su-30SM, and offers increased commonality with the Su-35S, including a new AL-41F-1S engine, plus full replacement of French-supplied avionics with Russian items. A Russian helmet-mounted sight replaces the Ukrainian-made Sura, more capable mission computers and enhanced radar are fitted as well as a multi-channel communication and tactical information distribution system. The Russian-standard Su-30SM has since been acquired by Kazakhstan (at least 24 ordered), Armenia (four, with more planned), and Belarus (four to date, 12 ordered).

The first customer for the Komsomolsk Su-30 was China, which acquired 76 Su-30MKK (the second K for Kitay, China) multirole fighters between 2000 and 2003, followed by 24 of the more advanced Su-30MK2 fighters in 2004 – the latter are operated by the navy. Compared to the Su-30MK2 version, the Su-30MKK has a simplified SUV-VE fire-control system with N001VE radar and no Kh-31A anti-ship missile capability.

China has since launched production of its indigenous J-16, which draws heavily upon these designs with locally produced avionics, weapons and engines. The next customer was Indonesia, which received two Su-30MKK fighters in 2003 and nine Su-30MK2 fighters between 2008 and 2013. Vietnam received 24 Su-30MK2V fighters between 2004 and 2012 and from 2014 began to receive another 12. In Russian parlance, the Vietnamese aircraft are actually designated Su-30 2 (the last letter being the Cyrillic B), while the Venezuelan aircraft are Su-30 2V, with the last letter written in Latin script. Venezuela received 24 Su-30MK2Vs between 2006 and 2008, while Uganda received six Su-30MK2s between 2011 and 2012.

Russia placed a first order for four Su-30M2 aircraft, featuring minor differences relating to some new software modes, national identification friend-or-foe (IFF), datalink, etc. in 2009 and 16 more in 2012. These have been delivered primarily to units flying the single-seat Su-27SM fighter for training purposes.

Air-to-air loads

All Su-30s can carry up to 8000kg (17,637lb) of stores on 12 hardpoints.

Su-30

Despite the obvious similarity of aircraft and colour scheme, this Su-30 of the Russian Knights was displaying 30 years later than the aircraft on page 34.

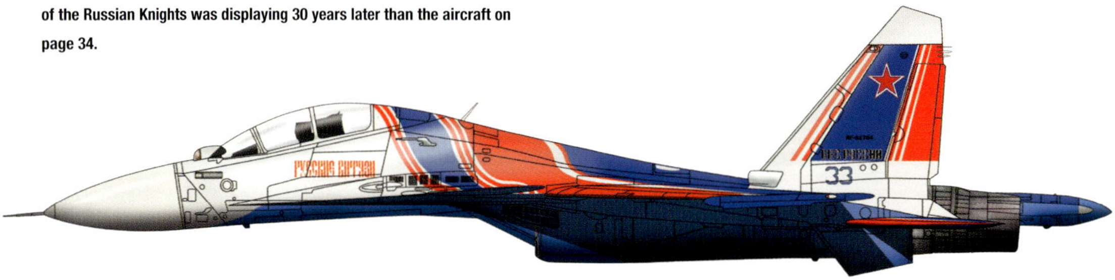

Air-to-air loads include up to eight semi-active radar-guided medium-/extended-range R-27R/ER (AA-10 Alamo) air-to-air missiles, two infrared-guided medium-/extended-range R-27T/ET air-to-air missiles, up to 10 active radar-guided medium-range R-77 (AA-12 Adder) and up to six infrared-guided short-range R-73 (AA-11 Archer) air-to-air missiles.

Air-to-surface options include two Kh-59M (AS-18 Kazoo) missiles, up to six Kh-31A/P (AS-17 Krypton) anti-ship/anti-radar missiles, up to six Kh-29L/T/TD (AS-14 Kedge) laser-/TV-guided missiles, up to three 1500kg (3307lb) or six 500kg (1102lb) guided bombs, up to 32 250kg (551lb) freefall bombs, plus unguided rockets. The Indo-Russian Brahmos-A air-to-surface missile is being integrated on the Su-30MKI.

As with all other Flanker variants, the Su-30 has one GSh-30-1 single-barrel 30mm (1.2in) cannon in the starboard leading-edge root extension.

Syrian war service

Russian Su-30SMs have been involved in combat operations over Syria since 2015, initially providing fighter escort for Russian attack jets and strategic bombers; the aircraft have also conducted their own air-to-ground strikes.

Sukhoi Su-33 NATO reporting name: Flanker-D

The carrier-capable Flanker was developed shortly before the collapse of the Soviet Union as a ship-based air superiority and air defence fighter.

A Russian naval Sukhoi Su-33 takes off at Zhukovsky airport after modernization with the 'Gefest' system.

The project to develop a carrier-capable Flanker dates back to the 1980s, forming part of plans to equip an air wing for the Soviet Navy's planned fleet of Project 1143.5 aircraft carriers, which would also have included the short take-off and vertical landing Yakovlev Yak-41 and the MiG-29K. Similar to the MiG-29K, the Su-27K (K for *Korabelnyi*, ship-borne) was developed to take off from a 'ski-jump' ramp. The first prototype, T-10K-1, made its maiden flight on 17 August 1987. This was followed by a first landing aboard the carrier *Tbilisi* on 1 November 1989.

Based on the Su-27 (described separately), the Su-33 differs most obviously from its land-based counterparts in featuring canard foreplanes to reduce the angle of attack for take-off and landing. The aircraft is also equipped with an arrester hook and strengthened undercarriage with twin nosewheels. Wing flaperons are larger, running the full length of the trailing edge, and full-span slats are fitted. Modifications appropriate to carrier use were made to the flight-control system and the aircraft features a greater proportion of anti-corrosion materials, including in its engines. For stowage on or below decks, a folding mechanism is incorporated in the outer wing panels, tailplanes, radome and tailboom. An inflight refuelling probe is included on the port side of the nose and provision for a buddy refuelling pod under the fuselage.

Up to 6500kg (14,330lb) of weapons and stores may be carried on 12 hardpoints. Air-to-air weapons are basically similar to the land-based Su-27, including medium-range and extended-range R-27 (AA-10 Alamo) air-to-air missiles and short-range R-73 (AA-11 Archer) air-to-air missiles, and the aircraft is fitted with the Flanker's standard GSh-30-1 single-barrel 30mm (1.2in) cannon.

Chinese export

Two prototypes (T-10K-1 and -2) were built by Sukhoi in Moscow, followed by seven pre-production aircraft (-T10K-3 to K-9) built at Komsomolsk-on-Amur in 1990 and 1991 used for tests. The Russian Navy received 26 production aircraft between 1993 and 1996. Two of the test aircraft (T-10K-3 and -7) were left in Ukraine after the collapse of the Soviet Union. A third, T-10K-7, was sold to China in 2004 and became the basis for China's J-15 ship-based fighter, subsequently further developed including the J-15D ship-based electronic warfare aircraft. The Su-27K was officially commissioned into service with the Russian Navy in August 1998 and received the revised designation Su-33. Currently, the Russian Navy has around 14 Su-33s within a single operational unit, the 279th Independent Ship-borne Fighter Aviation Regiment (OKIAP) based at Severomorsk-3 assigned to the Northern Fleet.

As part of their overhauls, the aircraft have undergone modest upgrades since 2009. The sole derivative to actually be built was the one-off Su-27KUB (*Uchebno-Boyevoi*, combat trainer), also known as the T-10KUB or Su-33UB, a side-by-side two-seat multirole fighter with Zhuk radar, that was first flown in April 1999 but later cancelled. Related unbuilt projects included the Su-27MK multirole fighter, Su-27KRTs reconnaissance and target indication aircraft, Su-27KPP electronic warfare aircraft and Su-33M mid-life upgrade with multirole capability added.

The first, and to date only, combat service was during the cruise by the *Admiral Kuznetsov* in the Mediterranean from November 2016 to January 2017 to participate in Syrian operations. The air group included four MiG-29KR/KUBR and nine Su-33 fighters, plus helicopters.

In December 2016, a Su-33 crashed into the Mediterranean after the arrester cable broke. Most missions were flown from Khmeimim air base in Syria.

Su-33

Weight (maximum take-off): 24,500kg (54,013lb)
Dimensions: Length 21.2m (69ft 6in) without probe, Wingspan 14.7m (48ft 3in), Height 5.72m (18ft 9in)
Powerplant: Two modified AL-31F series 3 turbofans, each rated at 122.5kN (27,558lb) with afterburning
Maximum speed: Mach 2.17
Range: 3000km (1864 miles)
Ceiling: 17,000m (55,775ft)
Crew: 1
Armament: One GSh-301 30mm (1.2in) cannon, plus up to 6500kg (14,330lb) of stores carried on 12 hardpoints

Su-33

Flying from Kubinka air base in 2016, '77 Red', like the much earlier aircraft depicted in the three-view opposite, wears the Sea Eagle badge of the 1st squadron of the Severomorsk Regiment on the tailfin.

Su-33

'64 Red' was one of the aircraft deployed aboard the carrier *Kuznetsov* during its initial cruise in 1996. The canards are a distinguishing feature of the Su-33 as no other single-seat 'Flanker' currently in service possesses these control surfaces.

Sukhoi Su-34 NATO reporting name: Fullback

Originally intended to fulfil air interdiction, reconnaissance and escort-jamming missions, the Su-34 is currently primarily used as a long-range strike aircraft, or a 'tactical bomber' in Russian parlance.

The first conception of a strike version of the Su-27 appeared in 1977 with the aim of developing a minimum-change version adding existing navigation and attack systems. However, it wasn't until 1986 that development of the Su-27IB (*Istrebitel-Bombardirovshchik*, fighter-bomber, or T-10V) was authorized.

Featuring avionics based around the new Sh141 air-to-ground radar system, the new aircraft was developed under the Furor (furore) programme. The first prototype T-10V-1, converted from a production Su-27UB, was built by Sukhoi in Moscow and flew on 13 April 1990. Subsequent production of Su-27IBs was handled by the Novosibirsk plant.

The Su-27IB fuselage was modified with a widened 'platypus' nose seating the pilot and the navigator/ weapons system operator side-by-side in a titanium-alloy armoured 'box and the tail' and the 'sting' was also considerably enlarged. Starting with the second prototype, T-10V-2, the landing gear was strengthened, with twin mainwheels. Mission avionics include the K-102 targeting and navigation complex that brings together a central computer, data display system, Platan laser/TV sight and navigation system. The Sh141 incorporates the V004 passive electronically scanned array radar, with air-to-ground and air-to-air modes, as well as the Khibiny electronic countermeasures (ECM) suite.

Upgraded prototypes

In 1994, the Su-27IB was officially redesignated as the Su-34. The last of the test aircraft, T-10V-8, took to the air in December 2003, at which point the Russian Air Force revised its requirements for the Su-34. Continued test work then made use of upgraded prototypes. In October 2006, the Su-34 was cleared for initial operational service, although during its trials phase test aircraft were involved in the conflict in Chechnya, being used to jam enemy air defence radars in 2000 and 2002. Two prototypes also took part in the war with Georgia in August 2008, again in an escort-jamming role. The 'second stage' of state evaluation lasted between 2006 and 2011 with the first operational regiment fully equipped with 24 new Su-34s by October 2013 at Voronezh.

Su-34

Weight (maximum take-off): 45,100kg (99,428lb)
Dimensions: Length 24.8m (81ft 4in), Wingspan 14.7m (48ft 3in), Height 6.08m (19ft 11in)
Powerplant: Two Saturn AL-31F turbofans, each rated at 122.58kN (27,558lb) thrust with afterburning
Maximum speed: Mach 1.6
Range: 4000km (2485 miles) ferry range
Ceiling: 15,700m (51,509ft)
Crew: 2
Armament: One GSh-301 30mm (1.2in) cannon, plus up to 8000kg (17,637lb) of external stores on 12 weapons pylons, including Kh-59M (AS-18 'Kazoo') TV-guided missiles, Kh-31P anti-radar and Kh-31A anti-ship missiles (AS-17 'Krypton'), TV- or laser-guided Kh-29T/L (AS-14 'Kedge') air-to-surface missiles and KAB-500 and KAB-1500 guided bombs

Su-34

With its radically redesigned fuselage, 'duck-billed' nose profile and enormous tail 'sting', the Su-34 is highly distinctive. This example was based at Baltimore air base near Voronezh in 2015.

Displaying during the MAKS Airshow at Zhukovsky in August 2015, the Su-34 has retained much of the startling agility of its Flanker forebears despite its maximum take off weight being nearly double that of an early Su-27.

The Su-34 can carry a total of up to 8000kg (17,637lb) of weapons and stores on 12 pylons: six under the wing, two under the engine intake ducts and two in tandem between the engines. The two wingtip stations usually are fitted with Khibiny ECM pods or otherwise R-73 (AA-12 Archer) air-to-air missiles for self-defence. The aircraft is able to carry the full range of Russian tactical air-to-surface missiles including the Kh-59M (AS-18 Kazoo) heavy TV-guided missile, and the supersonic Kh-31P anti-radar and Kh-31A anti-ship missiles (AS-17 Krypton). The Su-34 can also utilize other weapons including TV- or laser-guided Kh-29T/L (AS-14 Kedge) missiles and KAB-500 and KAB-1500 guided bombs. Satellite-guided munitions include the KAB-500S bomb and Kh-25MS missile. Air-to-air weapons include the beyond-visual-range R-27 (AA-10 Alamo) and R-77 (AA-12 Adder) missiles. Other options include unguided bombs, rockets and tactical nuclear bombs. More recent weapons to be integrated include the Kh-35U (AS-20 Kayak) subsonic anti-ship missile, the Kh-38M new-generation air-to-ground missile, and the 1500kg (3307lb) UPAB-1500 glide bomb. In addition, the aircraft mounts one GSh-30-1 single-barrel 30mm (1.2in) cannon in the starboard leading-edge root extension.

Wideband datalink

In its basic form, the Su-34 is a bomber intended primarily for tactical interdiction but originally a separate Su-34R reconnaissance version was planned with internal equipment, including a side-looking radar. This proved too expensive and the reconnaissance equipment is now provided in the form of external pods as part of the BKR-3 reconnaissance suite. The main sensor is the Pika

Su-27IB

'42 Blue' was the first flying prototype Su-27IB which first flew on 13 April 1990 and would subsequently be redesignated Su-34. A rebuilt Su-27UB with new radar and forward fuselage, the Su-27IB retained the single wheel main undercarriage of the standard 'Flanker'.

side-looking radar pod. A second electro-optical pod contains a TV camera and infrared line-scanner and a third pod contains a signals intelligence system. All three pods have a wideband datalink to transmit data to the ground in real-time. The escort jamming mission is carried out using the Khibiny ECM suite supplemented by four underwing jamming pods. The pods offer increased power and a wider frequency range than the basic ECM suite, allowing the aircraft to detect, classify and locate enemy radar before transmitting countermeasures. More recent additions are the Tarantul (tarantula) ECM escort jamming pods covering various frequency ranges.

A separate Su-34P escort jamming version had been planned with internal jamming equipment, but as with the reconnaissance variant this did not proceed to production.

The most recent variant, developed under the Sych (little owl) programme, the Su-34M is an upgrade of the original aircraft. It includes the improved Sh141M radar system and modernized K-102M nav/attack suite, as well as new types of weapons and stores. The first upgraded aircraft were scheduled to be in service before the end of 2021.

Since September 2015, Su-34s have been a regular presence in Syria. While primarily used to drop unguided ordnance, these aircraft have also employed precision weapons, including satellite-guided KAB-500S bombs and the Kh-35U anti-ship missile. More recently the Su-34 was committed to the Russian invasion of Ukraine, four of which were destroyed in the opening weeks of the conflict.

Su-27IB

Weight (maximum take-off): 45,100kg (99,428lb)

Dimensions: Length 24.8m (81ft 4in), Wingspan 14.7m (48ft 3in), Height 6.08m (19ft 11in)

Powerplant: Two Saturn AL-31F turbofans, each rated at 122.58kN (27,558lb) thrust with afterburning

Maximum speed: Mach 1.6

Range: 4000km (2485 miles) ferry range

Ceiling: 15,700m (51,509ft)

Crew: 2

Armament: One GSh-301 30mm (1.1in) cannon, plus up to 8000kg (17,637lb) of external stores on 12 weapons pylons, including Kh-59M (AS-18 'Kazoo') TV-guided missiles, Kh-31P anti-radar and Kh-31A anti-ship missiles (AS-17 'Krypton'), TV- or laser-guided Kh-29T/L (AS-14 'Kedge') air-to-surface missiles and KAB-500 and KAB-1500 guided bombs

Sukhoi Su-35 NATO reporting name: Flanker

A formidable single-seat heavy long-range multirole fighter developed from the Su-27, Russia has acquired the Su-35S version as a lower-cost supplement to the new-generation Su-57.

Soon after the service entry of the original Su-27, work began on testing a significantly improved Su-27M, with a first prototype completed during 1988 and flying for the first time on 28 June. In total, 12 evaluation aircraft were built, and in 1995 there followed a batch of three production aircraft. Meanwhile, the demise of the Soviet Union meant the Moscow-based Sukhoi Design Bureau began looking for export customers for what was now termed the Su-35.

Helmet-mounted sight

Compared to the Su-27, the Su-35 features a number of major changes. These initially included a revised aerodynamic configuration with canard foreplanes combined, in later prototypes, with thrust-vectoring engines and a quadruple-redundant digital fly-by-wire system for enhanced agility. Less obvious changes include lower tailfins with wider rudders, a shorter tail 'sting' containing an N012 rearward looking radar and the deletion of the dorsal airbrake. Internally, the aircraft makes much greater use of advanced composite materials in its structure. A modern Irbis passive electronically-scanned array radar, reportedly capable of detecting hostile aircraft at a range of 400km (882lb) was also introduced as part of a revised fire-control system. As well as the Irbis (or Irbis-E for export), the system includes the Khibiny-M electronic countermeasures suite. The Irbis is said to be capable of tracking-while-scanning up to 30 air targets, eight of which can be engaged simultaneously with active-radar missiles. As with other fighter Flankers, there is also an infrared search and track (IRST) sensor, the OLS-35, which combines thermal-imaging and TV cameras, a laser rangefinder and a target designator. The pilot has a Sura-M helmet-mounted sight.

The first iteration found no export customers, despite concerted marketing efforts in nations such as South Korea, Brazil, China and the United Arab Emirates. As a result, in the early 2000s Sukhoi launched the Su-35BM (BM for *Bolshaya Modernisatsiya*, literally 'big modernization', although later this was simply known as the Su-35). The maiden flight of the first prototype, Su-35-1, took place on 19 February 2008. The aircraft was then ordered for the Russian Air Force as the Su-35S in August 2009, with the first production taking to the air for the first time on 3 May 2011.

Sukhoi Su-35

Weight (maximum take-off): 34,500kg (76,059lb)

Dimensions: Length 21.9m (71ft 10in), Wingspan 14.7m (48ft 2in), Height 5.9m (19ft 4in)

Powerplant: Two thrust-vectoring Saturn AL-41F-1S (*izdeliye* 117S) turbofans, each rated at 86.29kN (19,400lbf) dry and 137.29kN (30,865lbf) with afterburning

Maximum speed: 2400km/h (1500mph)

Range: 3600km (2200 miles)

Ceiling: 10,000m (59,055ft)

Crew: 1

Armament: One 30mm (1.2in) Gryazev-Shipunov GSh-30-1 autocannon; up to 8000 kg (17637lb) of weapons and stores

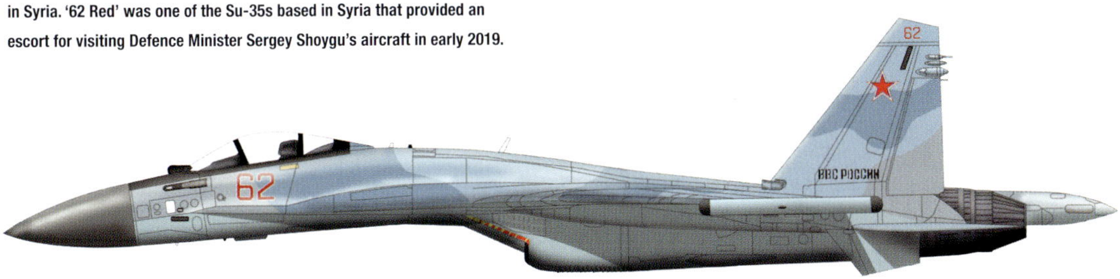

Su-35

The Su-35 has been active in Russia's ongoing involvement in the civil war in Syria. '62 Red' was one of the Su-35s based in Syria that provided an escort for visiting Defence Minister Sergey Shoygu's aircraft in early 2019.

Su-35

The second prototype of the Su-35, '902 Blue' was demonstrated in this striking splinter scheme at the MAKS Airshow, Moscow, in August 2011.

Su-35

On display at the Paris Airshow of 2013, '07 Red' was the first Su-35 to appear at an airshow outside Russia and demonstrate the type's thrust-vectoring super-manoeuvrability.

Su-35

Another star performer at airshows, '901 Blue' features a highly distinctive 'faceted' three colour camouflage scheme. This may be seen to good effect in the photograph on page 8 at the start of this chapter.

Sukhoi Su-35

Weight (maximum take-off): 34,500kg (76,059lb)
Dimensions: Length 21.9m (71ft 10in), Wingspan 14.7m (48ft 2in), Height 5.9m (19ft 4in)
Powerplant: Two thrust-vectoring Saturn AL-41F-1S (*izdeliye* 117S) turbofans, each rated at 86.29kN (19,400lbf) dry and 137.29kN (30,865lbf) with afterburning
Maximum speed: 2400km/h (1500mph)
Range: 3600km (2200 miles)
Ceiling: 18,000m (59,055ft)
Crew: 1
Armament: One 30mm (1.2in) Gryazev-Shipunov GSh-30-1 autocannon; up to 8000 kg (17637lb) of weapons and stores

Su-35S

Russia initially ordered 48 Su-35S fighters with plans for another 48 Su-35S to be ordered subsequently. The most obvious external change from the earlier Su-35 airframe was the deletion of the canard foreplanes, Sukhoi having concluded that the manoeuvrability lost with the removal of these control surfaces could be more than made up for with the thrust vectoring nozzles while dispensing the weight penalty imposed by the canards. As a result, the Su-35S more closely resembles the Su-27 from which it was originally developed than the earlier iteration of the Su-35 design.

Air-to-ground weapons

The Su-35S can carry up to 8000kg (17,637lb) of weapons and stores on its 12 hardpoints. All types of air-launched tactical missile currently in the Russian inventory can be employed and are due to be joined by new weapons in the future. Compared to other Flanker variants, weapons available to the Su-35S include the very long-range R-37M (AA-13 Axehead) air-to-air missile. This is complemented by the medium-range R-77-1 and short-range R-74M missiles, which represent the latest developments of the R-77 (AA-

Landing after its display at MAKS 2015, '03 Red' is operated by the 4th Centre for Combat Application and Crew Training (GTsPAPIPVI) based at Lipetsk. This unit displays four Su-35s as the 'Falcons of Russia' combat demonstration team.

12 Adder) and R-73 (AA-11 Archer) families respectively. Air-to-ground weapons include the Kh-31PM and Kh-58USh anti-radiation missiles; the Kh-31AM, Kh-35U and Kh-59MK anti-ship missiles; and the 'universal' family of Kh-38M missiles that can be fitted with various seekers. Like its predecessor, the Su-35 carries one GSh-30-1 single-barrel 30mm (1.2in) cannon in the starboard leading-edge root extension.

As of the mid-2000s it was expected the Su-35S would be ready for service fairly quickly and would serve as a stopgap fighter pending availability of what would become the Su-57. In the event, the new-generation fighter programme suffered considerable delays and the Su-35S is now viewed more as a lower-end complement to the more-costly Su-57 rather than an aircraft that the Su-57 will replace.

Russian Aerospace Forces

The first Su-35S was handed over to the then Russian Air Force (now the Russian Aerospace Forces) at the end of 2011. In February 2014 the first batch of 12 Su-35S fighters was delivered from Sukhoi's Komsomolsk-on-Amur Aircraft Plant to the first operational Su-35 air base at Dzyomgi in the Russian Far East. By 2015, a total of 10 aircraft had been handed over to flight test centres and 24 were based at Dzyomgi, while the second and third operational air bases at Tsentralnaya Uglovaya and Byesovyets had received 10 more between them. By the end of 2018 another 26 examples had been delivered to Russian units for an overall total of 78 aircraft. By this point in time, the Komsomolsk-on-Amur factory had reportedly constructed 100 serial-production airframes.

After the successful service introduction of the Su-35S in Russia, orders for the export Su-35 version were placed by China and Egypt. The 24 examples supplied to China currently serve with the 6th Air Brigade at Suixi. China received its first four aircraft in December 2016, with a further two batches of 10 each being delivered during 2017 and 2018. Egypt has reportedly ordered between 24 and 31 examples of the type, of which at least five have been manufactured. However, as of early 2022 there was no sign that these had been delivered, with trials activity continuing in Russia.

Since September 2015, the Russian Aerospace Forces contingent in Syria has been used to support the regime in the ongoing civil war. Initially, Sukhoi's Su-30SM and Su-27SM fighters were deployed, but since January 2016 Su-35s have been flown there and used for high-profile escort missions of long-range bombers, although it is believed that Su-35s have also performed bombing missions of their own. The aircraft has reportedly intercepted Turkish and Israeli F-16s on several occasions and turned them away from Syrian airspace, although these reports remain unconfirmed.

Sukhoi Su-57 NATO reporting name: Felon

A fifth-generation single-seat heavy long-range multirole fighter, the Su-57 features low observability, supersonic cruising speed (supercruise) and supersonic manoeuvrability.

In April 2001, the Russian Air Force announced a competition for the PAK FA (*Perspektivnyi Aviatsionnyi Kompleks Frontovoi Aviatsii*, or Future Air Complex of Tactical Aviation). The competitors were the Sukhoi T-50 and the smaller and cheaper Mikoyan E-721. The government selected the T-50 design for the PAK FA project, and in July 2003, Sukhoi received a formal development contract from the Russian Ministry of Defence to include design, construction and testing of five T-50 prototypes. In the same year a letter of intent was signed by Russia and India covering common development, India's T-50-based PMF (Prospective Multirole Fighter, Type 79L), being intended for that country's Fifth-Generation Fighter Aircraft (FGFA) programme.

The first prototype T-50-1 performed its maiden flight in Komsomolsk-on-Amur on 29 January 2010, later moving to Zhukovsky for more extensive trials. Unfortunately, the prototypes were found to have problems related to airframe strength, demanding a major reworking of the design. T-50-1 suffered a structural crack that needed over a year to repair and T-50-5 caught fire on the runway at Zhukovsky in June 2014. After rebuild, it emerged as the T-50-5R, resuming flight tests in October 2015.

State acceptance trials

In 2012, the decision was made to build another batch of test aircraft according to a revised and strengthened design. The first of these, T-50-6, made its maiden flight on 27 April 2016, followed by four further aircraft, T-50-8 to T-50-11, the last two of which were fully equipped with mission sensors and avionics. In July 2017, the T-50 officially received the Su-57 designation with state acceptance trials completed in May 2018. The State Armament Programme for 2011 to 2020 had called for acquisition of 52 PAK FAs by 2020, with 150-160 more in subsequent years. However, the programme was rocked by India's withdrawal in 2018, as well as Western sanctions on Russia and an economic downturn. In 2018, the United Aircraft Corporation (UAC) was awarded a contract for just two fighters to be delivered to the Russian Aerospace Forces, one in 2019 and a second in 2020. By December 2019, the Komsomolsk-on-Amur plant had completed the first pre-production Su-57 (T-50S-1) but the aircraft crashed during its handover flight. The second aircraft, T-50S-2, was completed in October 2020. This was the first to be handed over to the Russian Aerospace Forces in December 2021. In May 2019 Putin called for the purchase of 76 Su-57 fighters by 2028.

Su-57

Weight (maximum take-off): 35,000kg (77,162lb)

Dimensions: Length 20.1m (66ft), Wingspan 14.1m (46ft 3in), Height 4.6m (15ft 1in)

Powerplant: Two Saturn AL-41F1 (*izdeliye* 117) thrust-vectoring turbofans, each rated at 147.2kN (33,100lb) thrust with afterburning

Maximum speed: Mach 2.0

Range: 3500km (2175 miles)

Ceiling: 20,000m (65,617ft)

Crew: 1

Armament: One GSh-301 30mm (1.2in) cannon, plus disposable ordnance carried in two tandem weapons bays in the lower fuselage, typically up to four R-77M AAMs, Kh-58UShK or Kh-36 anti-radiation missiles, or Kh-38M or KAB-250 precision-guided munitions. Two R-74M2 AAMs carried in weapons bays in underwing fairings, plus option for external ordnance on four underwing pylons.

Su-57

Despite bearing the serial number 01, this is actually the second full production machine, which was renumbered following the delivery flight crash of the first example in December 2019. The aircraft wears the badge of the 929th Chkalov State Flight-Test Centre with which it serves.

T-50-9

The ninth prototype, and eighth flyable airframe of the Su-57, also known as the T-50-9. First flown on 24 April 2017, the aircraft wears a pixelated camouflage pattern.

Series production

By February 2022, the Russian Ministry of Defence had taken delivery of just three production Su-57s (T-50S-2, S-3 and S-4). The first batch of Su-57s is expected to be delivered to an operational unit during 2022 and 22 fighters are planned to be delivered by 2024. An order for the modernized Su-57M (T-50M), also known as the 'second stage' Su-57, was placed in October 2018 with series production planned by the end of 2024. The major change is the new *izdeliye* 30 turbofan engine, replacing the current AL-41F-1, with increased thrust, lighter weight and lower operating costs.

The Su-57's aerodynamic configuration is optimized for high lift-to-drag ratio at supersonic speed. The airframe includes fuselage side extensions that terminate in large moving flaps on the leading edge. These increase static instability for high-speed manoeuvrability. Fire-control systems include the Sh121 radio-electronic suite and 101KS electro-optical suite; both also contribute to self-protection. Sh121 consists of the N036 radar system, N036Sh identification friend or foe (IFF), and L402 electronic countermeasures (ECM). N036 radar has five active electronically scanned arrays (AESA). The 101KS consists of an infrared search-and-track (IRST) sensor ahead of the cockpit, four ultraviolet missile-approach warning sensors, two directional infrared countermeasures turrets and one imaging infrared sensor for low-level flying. Unlike previous Russian tactical fighters, the Su-57 had a navigation and targeting pod developed for it from the outset.

The main ordnance load is carried in two large internal weapon bays, arranged in tandem between the engines. New air-to-air missiles have been developed for internal carriage, including the beyond-visual range R-77M and the very long range (around 300km/186 miles) *izdeliye* 810. Two additional missiles are contained in 'quick launch' bays in underwing fairings; each of these bays carries one R-74M2 infrared-guided air-to-air missile. Air-to-surface weapons designed specifically for the Su-57 include the Kh-69 standoff missile, with a range of 300km (186 miles) or more. The Kh-58UShK anti-radiation missile possesses folding wings and fins to fit in the internal bays. Other options include the Kh-38M air-to-ground missile (semi-active laser, imaging infrared or active-radar seeker options). Smaller air-to-ground munitions include the 250kg (551lb) KAB-250L electro-optical guided bomb. Gun armament is provided by one GSh-301 (9A1-4071K) 30mm (1.2in) single-barrel cannon in the starboard wing root.

In February 2018, two Su-57s (T-50-9 and T-50-11) deployed to Syria for a few days for combat trials. Reportedly, the aircraft used the new Kh-59MK2 heavy air-to-surface missiles while stationed there.

Viewed from underneath, the Su-57's distinctive planform becomes apparent. With only a handful in service, it remains to be seen if the 'Felon' will prove to be as successful as the 'Flanker' in its country of origin and on the export market.

Sukhoi Checkmate

An all-new single-engine fighter, as yet unflown, the Checkmate is optimized for low-visibility, supersonic speed and agility.

Known as the Light Tactical Aircraft, or LTS in Russian, the new Checkmate is primarily marketed as an export product. Moscow had almost given up on purchases of lightweight and mediumweight fighters in favour of the Flanker family and, more recently, the Su-57. Compared to the Su-57, the Checkmate features a similar blended-body configuration with a lift-generating fuselage, and wing leading-edge root extensions.

The engine is rumoured to be the *izdeliye* 30 planned for the definitive Su-57, although the published thrust figures of 14,500–16,000kg (31,967–35,274lb) for the Checkmate would suggest a lower-rated engine than the *izdeliye* 30. The aircraft has a planned weapons load of 7400kg (16,314lb), with stores carried on five internal hardpoints (located within three weapons bays) as well as six more pylons located externally.

International air show

Planning for a new single-engine fighter reportedly began in 2016 after studying Russian experiences in the Syrian conflict, and rumours of a new Russian fighter design emerged in July 2021 when the United Aircraft Corporation (UAC) posted a cryptic teaser on its Twitter account. A presumed mock-up of the Checkmate was spotted the same month at Zhukovsky wrapped in a black protective covering. It was then formally unveiled at MAKS international air show in Moscow the same month.

In November 2021 it was announced that the first flying prototypes were under construction. Russian media stated that United Arab Emirates could become involved in the programme and have also pointed to the potential of selling the fighter to the Russian military. Other potential customers include countries in the Middle East, Asia Pacific region and Latin America.

UAC claims a unit price of just $30 million compared to around $90 million for the American-built Lockheed Martin F-35 Lightning II. Industry officials have spoken of a potential two-seater as well as pilot-optional and unmanned Checkmate variants.

Sukhoi Checkmate
Weight (maximum take-off): 18,000kg (39,680lb)
Dimensions: Length 17.4m (57ft), Wingspan 11.9m (39ft)
Powerplant: Rumoured to be a variant of izdeliye 30, published thrust figures of 142.2kN (31970lbf) to 156.91kN (35274lbf)
Maximum speed: 2200km/h (1370mph)
Range: 2800km (1740 miles)
Crew: 1
Armament: Up to 7400kg (16,314lb) of weapons and stores in internal bays and external hardpoints

Sukhoi Checkmate
As yet unflown, the Checkmate currently exists only in mock-up form. While a lower cost alternative to the F-35 would appear to be an attractive proposition for many nations, it is impossible to say with any certainty whether this will be borne out in reality.

BOMBERS

Historically, the Soviet Air Force was primarily a tactical force, but the strategic arm of its operations was the most well-known branch of its operations in the West, NATO fighters regularly intercepting Tupolev Tu-95 'Bears' and later Tupolev Tu-22M 'Blinders' and Tupolev Tu-160 'Blackjacks' as they probed the edges of international air space. The modern Russian strategic bomber fleet continues that tradition.

This chapter includes the following aircraft:

- Tupolev Tu-22M
- Tupolev Tu-95MS
- Tupolev Tu-160

As emblematic a symbol of long range Cold War aviation as the United States' B-52, the Tu-95 'Bear' still serves with Russian aerospace forces some 70 years after its first flight (which coincidentally took place in the same year as the B-52). With its propeller tips exceeding the speed of sound, the Tu-95 is one of the loudest military aircraft ever built. These Tu-95SMs feature underwing cruise missile pylons between the engines.

Tupolev Tu-22M NATO reporting name: Backfire

A supersonic intermediate-range bomber and missile carrier intended to deliver both nuclear and conventional weapons, the Tu-22M has been in service for over 40 years.

The Tu-22 was developed by the Tupolev Design Bureau in Moscow and manufactured by the Kazan Aircraft Plant. Both the Tupolev Design Bureau and the Kazan Aircraft Plant are today part of the Tupolev Company.

After the disappointing performance of the original Tu-22 (NATO reporting name: Blinder), the Soviet government called upon Tupolev to develop the Tu-22M in November 1967. Although maintaining the Tu-22 designation, the Tu-22M was effectively a completely new design and was to prove vastly more successful than its predecessor. Requirements included a maximum speed of 2300-2500km/h (1429–1553 mph) and subsonic range of 7000km (4349 miles) armed with one Kh-22 missile.

Reconnaissance aircraft

A batch of eight pre-production Tu-22M0 Backfire-A was constructed, the first of which took to the air on 30 August 1969. Unfortunately, the Tu-22M0 failed to meet the performance criteria so it was followed by the Tu-22M1 that appeared in 1971, with new movable outer wing panels, but only nine were produced. The first major production version was the Tu-22M2 Backfire-B featuring a slightly revised airframe that appeared in 1973, of which 211 were built until 1983. However, the Tu-22M2 still did not meet the original performance expectations, so the design was further developed into the definitive Tu-22M3 Backfire-C. Featuring wings that can be folded back to 65 degrees, instead of 60 degrees, more powerful NK-25 engines and wedge-type air intakes, the Tu-22M3 is faster than earlier variants and thanks to reduced fuel consumption it possesses greater range.

First flown in 1977, 268 Tu-22M3s were built until production ceased in 1993. In addition, a dozen examples of the Tu-22M3 were completed as Tu-22MR reconnaissance aircraft. This has side-looking radar, an electronic intelligence system, infrared scanner and various optical cameras.

Tupolev Tu-22M3 Backfire

Weight (maximum take-off): 124,000kg (273,373lb)

Dimensions: Length 42.46m (139ft 4in), Wingspan 23.3m (76ft 5in) fully swept or 34.28m (112ft 6in) fully spread, Height 11.05m (36ft 3in)

Powerplant: Two Kuznetsov NK-25 (*izdeliye Ye*) after-burning turbofans, each rated at 140.2kN (31,526lbf) dry and 245.18kN (55,115lbf) with afterburning

Maximum speed: 2300km/h (1429mph)

Range: 6800km (4225 miles)

Ceiling: 14,000m (45,932ft)

Crew: 4

Armament: One 23mm (0.906in) Gryazev-Shipunov GSh-23 twin-barrel Gast autocannon in remotely controlled tail turret; up to 24,000kg (53,000lb) of bombs, missiles or mines

Tu-22M3

With its wings set at minimum sweep angle for cruise, Tu-22M3 '58 Red' is shown with a Kh-22 'Kitchen' anti-ship missile carried under the port wing glove.

Cold War roles included anti-shipping strikes against US Navy aircraft carriers and other major warships, especially in the Mediterranean and the Atlantic Ocean, as well as interdicting convoys bringing reinforcements and supplies from the US to Europe in time of conflict. As part of the same mission, Tu-22Ms would have targeted European ports and larger airfields.

Production of all versions totalled 514. Although the aircraft was never exported, the collapse of the Soviet Union left Tu-22Ms in service in Russia and Ukraine. Ukrainian Tu-22Ms had all been scrapped by 2006. Today, the Tu-22M3 and Tu-22MR remain in service with the Russian Air Force only, after transfer of Russian Navy aircraft in 2011. Current operational bases are Belaya, Olenyegorsk and Shaykovka, with examples based at Ryazan for training. Around 60 Tu-22Ms remain in service with more in storage. Main armament is the Raduga Kh-22 (AS-4 Kitchen) supersonic heavy missile that dives onto its target at a speed of Mach 4.15. This missile is available with an active radar seeker version for use against ships or with inertial guidance for use against large ground targets.

Tupolev Tu-22M Backfire

The Tu-22 remains the most commonplace of Russia's large bombing aircraft and has been in service for 50 years. Its size and speed remain impressive and a series of upgrades have resulted in the Tu-22 remaining a highly capable combat aircraft.

Tupolev Tu-22M3 Backfire

Weight (maximum take-off): 124,000kg (273,373lb)

Dimensions: Length 42.46m (139ft 4in), Wingspan 23.3m (76ft 5in) fully swept or 34.28m (112ft 6in) fully spread, Height 11.05m (36ft 3in)

Powerplant: Two Kuznetsov NK-25 (*izdeliye Ye*) afterburning turbofans, each rated at 140.2kN (31,526lbf) dry and 245.18kN (55,115lbf) with afterburning

Maximum speed: 2300km/h (1429mph)

Range: 6800km (4225 miles)

Ceiling: 14,000m (45,932ft)

Crew: 4

Armament: One 23mm (0.906in) Gryazev-Shipunov GSh-23 twin-barrel Gast autocannon in remotely controlled tail turret; up to 24,000kg (53,000lb) of bombs, missiles or mines

Swing wings
The large variable geometry wings are swept forward as shown here for efficient cruising flight and swept back for a high speed dash over the target.

Engine
The Tu-22M3 is powered by two Kuznetsov NK-25 turbofan engines, one of the most powerful supersonic jet engines in service in the world today.

Tu-22M3

RF-34079, '57 Red', seen here with its wings at maximum sweep, displays its unusual shark-mouth markings applied to the engine intakes.

Nuclear capability

The anti-ship version is available in both nuclear and conventional form, but the land attack version is nuclear only. Closely resembling the Kh-22, the Kh-32 is a much-modernized version that is in use alongside the older weapon. Up to three of these heavy missiles can be carried by the Tu-22M, but a normal operational load is one or two. Freefall bombs can also be carried inside the bomb bay as well as on four external multiple racks under the engine air intakes and under the wings. The tail turret is equipped with a 23mm (0.9in) GSh-23M twin-barrel cannon with radar sight and is mainly used to fire countermeasure rounds.

Some Tu-22M3s have received the SVP-24-22 upgrade, based on a similar upgrade for the Su-24M. It adds a new computer and navigation system, plus new displays for the navigator and imparts increased navigational and improved bomb-aiming accuracy.

More extensive is the Tu-22M3M upgrade that includes new radar, as well as a new navigation system, communication suite, IFF and digital flight control system. This aircraft also boasts a partial glass cockpit and an improved self-defence suite. The first updated Tu-22M3M was flown in December 2018 with deliveries due to start in 2021.

Combat debut

First used in combat in Afghanistan in 1984, the Tu-22M was used to drop freefall bombs, a task it repeated in Chechnya during the mid-1990s and the war with Georgia in 2008. During the latter conflict one example was shot down by a surface-to-air missile. Since 2015, the aircraft has been employed in the Syria campaign, again dropping freefall bombs. In 2021, aircraft were deployed to Syria and flew patrol missions armed with Kh-22/Kh-32 anti-ship missiles.

Photographed at the moment of touchdown, Tu-22M3 '24 Red' makes a tyre-smoking arrival at Dyagilevo air base, Ryazan Oblast, Russia.

Tupolev Tu-95MS NATO reporting name: Bear-H

Arguably the single most iconic Soviet aircraft of the Cold War era, the slender 'Bear' has been in service since 1956 and is forecast to continue flying with Russian Aerospace Forces until at least 2040.

Despite often being labelled a strategic bomber, the intercontinental-range Tu-95MS currently in service exclusively operates with a missile armament and is more accurately a missile carrier. Regular armament include six Kh-55 (on Osina aircraft) or Kh-55SM (Sprut aircraft) long-range nuclear cruise missiles carried on a rotary launcher inside the weapons bay. These missiles are known to NATO as the AS-15 Kent.

The Tupolev Design Bureau, still led at the time by Andrei Tupolev himself, was invited to submit a design for an intercontinental bomber with an unrefuelled range of 8000km (4971 miles) in 1950. Tupolev proposed a turboprop powered design with a wing swept at 35 degrees, as this combination offered the best combination of speed and range then possible. The design was approved by the Government in July 1951 as the Tu-20, although by the time aircraft started to arrive at operational units the designation had changed to Tu-95. Despite this, US and NATO intelligence continued to refer to the aircraft as the Tu-20 for many years.

Design variations

The aircraft was first flown in prototype form on 12 November 1952, with series production being launched during 1955 in Samara and continuing until 1969 with a total of 173 built. In addition to the original Tu-95M bomber version, there were also Tu-95K missile carriers carrying the Kh-20 (AS-3) cruise missile that entered service in 1960 and the Tu-95RTs maritime reconnaissance aircraft, which was equipping naval aviation units from the mid-1960s. Production then switched to the Tu-142, a dedicated maritime patrol aircraft based on the Tu-95, which was built in Taganrog. Two examples of an airliner version, the Tu-116, were also built, although this was merely a back-up in case of failure of the more comprehensively modified Tu-114 airliner that mated a new fuselage to the wings, undercarriage and tail unit of the Tu-95.

In the late 1970s, attention switched to employing the Tu-95 solely as a missile-carrier with new long-range cruise missiles as its offensive armament. A Tu-95M was converted as a prototype and first flown in this form on 31 July 1978. Moscow decided to procure a new missile-carrier – Tu-95 – to complement the all-new Tu-160. The new Tu-95 variant would be based on the Tu-142M maritime patrol aircraft, as this was the most modern of the Bear 'family' at the time. The forward fuselage was reduced in length, the

This Russian Air Force Tu-95MS, '62 Red', was photographed at Zhukovsky International airport near Moscow during October 2014. Operational Tu-95s are based at Engels-2 air base in Western Russia and Ukrainka air base in the Far East.

flight deck widened and the aircraft first flown on 14 September 1979.

Bear-H6

Launched into production at the Samara plant as the Tu-95MS in 1981, first deliveries were made to the Soviet Air Force in December 1982. Two main versions have been fielded, the first equipped with an Osina missile control system for the Kh-55 missile and, from 1986, a later version with the more modern Sprut system compatible with Kh-55SM missiles and featuring improved defensive systems. While the Kh-55SM is able to carry additional fuel tanks for extended range, these cannot be fitted when the missile is carried by the Bear. Some Tu-95MS only carried six missiles internally and are sometimes known as the Tu-95MS6 or Bear-H6, while the Tu-95MS16 or Bear-H16 adds provision for another 10 missiles under the wing, two missiles fitted on each inboard pylon and three missiles on each pylon between the engines. As defensive armament, the Tu-95MS, like its predecessors, featured a manned tail turret equipped with two Gryazev-Shipunov GSh-23 23mm (0.9in) autocannon.

Distribution and current service

A total of 88 Tu-95MS were produced during 1982–92. After the demise of the Soviet Union, examples of the aircraft were left in Kazakhstan, Russia and Ukraine. Kazakhstan returned its aircraft to Russia in the mid-1990s and Ukraine sold its Sprut-equipped aircraft to Russia in 1999.

Beginning in 2003, the Tu-95MS added the Kh-555 (AS-22 Kluge) missile to its armoury, with six carried on the internal rotary launcher. These missiles are conventionally armed conversions of the Kh-55. Upgraded aircraft can carry six examples of the non-nuclear Kh-555 internally, as well as the stealthy Kh-101 and Kh-102 cruise missiles, of which eight can be carried in the bomb bay.

More recently, the Sprut-equipped aircraft have added stealthy Kh-101 (AS-23A Kodiak; conventionally armed) and Kh-102 (AS-23B; nuclear

armed) cruise missiles. These are too long to be contained within the internal weapons bay and are carried on four underwing pylons, each with two missiles. Adaptations for the new missiles began in 2015.

Currently, Russia has around 50 Tu-95MS in service of both the Osina and Sprut versions, and around half the fleet are Tu-95MS16 aircraft. Operational air bases are at Engels and Ukrainka, plus the training centre at Ryazan and a handful of test aircraft at Zhukovsky. As a separate effort to the weapons upgrades, aircraft are also being brought up to an improved standard to be known as Tu-95MSM, work on which was launched in 2009. Improvements are being introduced in stages, starting with new navigation equipment (satellite navigation receiver, VOR/ILS, short-range radio navigation system, automatic direction finder, radar altimeters, etc.).

Series upgrades began in 2014, with a handful of aircraft being

Tu-95MS

Named 'Veliki Novgorod', in honour of the city of the same name in the far west of Russia, RF-94124 '16 Red' is one of the Tu-95MS bombers operated by the 184th Heavy Bomber Aviation Regiment at Engels air base. Many examples of the Tu-95 fleet have been individually named.

Tu-95MS

Tu-95MS RF-94117 '27 Red' is named 'Izborsk' after the Medieval fortress complex of that name in Western Russia. 'Izborsk' is also on the strength of the 184th Heavy Bomber Aviation Regiment.

processed each year. The next stage of the Tu-95MSM upgrade involves adding a new Novella radar as well as a new flight control system and definitive navigation suite (inertial navigation, astro navigation and navigation computer).

The cockpit also receives five digital displays and the self-defence suite is updated. In addition, the engines of the Tu-95MSM have been upgraded to the NK-12MPM and fitted with new AV-60T propellers, providing better efficiency. The first flight of a Tu-95MSM was in August 2020.

Syrian combat

Despite having been in service for nearly 60 years at the time, the first use of the Tu-95 in combat eventually occurred over Syria on 17 November 2015 when three aircraft hit targets with eight Kh-555 cruise missiles. The aircraft flew direct from Engels air base, routing over the Caspian Sea and launched their missiles over Iran.

Tupolev Tu-95MS Bear

Weight (maximum take-off): 185,000kg (407,885lb)

Dimensions: Length 49.13m (161ft 2in), Wingspan 50.04m (164ft 2in), Height 13.30m (43ft 8in)

Powerplant: Four 11,185ekW (15,000ehp) Kuznetsov/Samara NK-12MP turboprops

Maximum speed: 830km/h (516mph)

Range: 10,500km (6524 miles)

Ceiling: 10,500m (34,450ft)

Crew: 6 or 7

Armament: Two 23mm (0.906in) Gryazev-Shipunov GSh-23 autocannon in tail turret; up to 15,000kg (33,000lb) bombload, including cruise missiles mounted on underwing pylons

Tupolev Tu-160 NATO reporting name: Blackjack

Capable of exceeding Mach 2, the spectacular Tu-160 is the world's largest and heaviest combat aircraft as well as being the largest variable geometry aircraft yet built.

An intercontinental-range missile carrier and bomber, carrying almost exclusively sub-sonic cruise missiles with nuclear or conventional warheads, the Tu-160 is intended to be able to penetrate enemy airspace at high speed at either high or low altitude. Originally the Tu-160 carried only missile armament but has reportedly been adapted to carry some freefall precision-guided ordnance.

The Tu-160 programme was launched in 1967 as the Tupolev Design Bureau's re-sponse to a competition for a supersonic strategic bomber with a range of up to 18,000km (11,18 miles) carrying the very large supersonic Kh-45 or Kh-2000 missiles. With the cancellation of these weapons, the armament was replaced with subsonic cruise missiles, the same Kh-55 (AS-15 Kent) as selected for the Tu-95MS. The first prototype took to the air on 18 December 1981 and the Tu-160 subsequently set 44 world records in its weight class between 1989 and 1990.

Tupolev completed two flying and one static prototype in Moscow, followed by 33 more aircraft (one for static test) built in Kazan. The first aircraft delivered went to Pryluky, Ukrainian SSR in 1987, which was the only operational Tu-160 base during the Soviet era. Ukraine would inherit 19 Tu-160s after the demise of the USSR, although by 1999 eight of these were transferred to Russia, with another 10 scrapped and one placed in a museum.

Six more Russian aircraft were constructed and delivered to the new Engels base in Russia during the early 1990s, but during 1994 production was terminated, leaving four unfinished aircraft at the Kazan factory. Of those four, two were finally completed and delivered in 1999 and 2007 respectively. To boost the Russian fleet, an aircraft that had been retained by Tupolev for test work was also transferred to the air force in 2006. The current Russian Tu-160 fleet consists of 16 aircraft based at Engels, with more undergoing overhaul and upgrade at Kazan. A further example is assigned for test work at Zhukovsky.

Nuclear capability

The Tu-160 retained the large fuselage tandem weapons bays, originally planned to accommodate Kh-45 missiles. Initially, the armament comprised a six-round rotary launcher in each bay, for a maximum total of 12 Kh-55SM cruise missiles with nuclear warheads. As an alternative armament, up to 24 Kh-15 (AS-16 Kickback) short-range attack missiles could be carried, although these have apparently been withdrawn from use.

In the early 2000s, the aircraft was provided with non-nuclear Kh-555 missiles that were first tested by operational Tu-160s over the Kola peninsula in 2005, with President Vladimir Putin flying in one of the aircraft. Beginning in the early 2010s, the stealthy Kh-101 (AS-23A Kodiak; conventionally armed) and Kh-102 (AS-23B; nuclear armed) cruise missiles were introduced. Up to 12 can be

Tu-160

RF-94100 '10 Red' is named 'Nikolai Kuznetsov' and is on the strength of the 121st Heavy Bomber Aviation Regiment, the sole operational unit flying the 'Blackjack' today. The Tu-160 fleet is based at Engels-2 air base in Saratov Oblast.

Tu-160

Completed using an unfinished airframe at the Kazan plant and first flown in December 2007, RF-94115 '08 Red' of the 121st Heavy Bomber Aviation Regiment is named 'Vitali Kopylov'. It was the newest aircraft in the Tu-160 fleet until the first new-build Tu-160 flew in January 2022.

Tupolev Tu-160 Blackjack
Weight (maximum take-off): 275,000kg (606,270lb)
Dimensions: Length 54.1m (177ft 6in), Wingspan 35.6m (116ft 9in) fully swept or 55.7m (182ft 9in) fully spread, Height 13.1m (44ft)
Powerplant: Four Kuznetsov NK-32 (*izdeliye R*) turbofans, each rated at 137.3kN (30,865lb) dry and 245.18kN (55,115lbf) with afterburning
Maximum speed: 2000km/h (1243mph)
Range: 13,950km (8668 miles)
Ceiling: 15,600m (51,181ft)
Crew: 4
Armament: Up to 45,000kg (99,208lb) of bombs or missiles

carried, on a pair of new six-round rotary launchers. Reports suggest that 1500kg (3307lb) UPAB-1500 glide bombs can also be carried.

Modernization programme

Over the 2010s, the Tu-160 has been subject to a thorough upgrading programme. Initially a comprehensive mid-life modernization was launched, the 'deeply modernized' Tu-160M. Approved in 2014, it includes a new Novella radar and a glass cockpit.

The revised navigation suite includes new inertial navigation, astro navigation, navigation computer, navigation radar, satellite navigation receiver, autopilot, and communication suite. Also new are the self-defence suite and reconnaissance systems. The first Tu-160M took to the air in 'deep modernized' form in February 2020.

In addition to this programme, a more modest 'first-stage' upgrade was implemented beforehand as the Tu-160M1. This does not include the more expensive elements such as the new radar and glass cockpit but features new autopilot, navigation complex and communication systems. A prototype was completed in 2006, followed by series upgrades in Kazan with aircraft returning to service during 2014. In 2015 Moscow also launched a plan to restart production of the Tu-160, pending the appearance of the all-new PAK DA strategic bomber as the Tu-160M2. However, the new-production version is now also known as the Tu-160M, reflecting the fact that it incorporates the same new systems as the Tu-160M upgrade.

The plan also required restarting production of the NK-32 engine, now in improved Series 02 form featuring improved reliability, reduced fuel consumption and greater lifespan. First flight of the new aircraft occurred on 12 January 2022 and Russian Aerospace Forces have specified a requirement for at least 50 new Tu-160Ms.

First combat role

The Tu-160 was first used in combat on 17 November 2015 to attack Syrian targets in concert with other bombing aircraft, flying direct from their base at Engels – although at least one mission was flown on a circular path from the Kola peninsula around Norway and the UK to release missiles over the Mediterranean before returning to Engels, a round trip of some 13,000km (8070 miles).

HELICOPTERS

The Russian military can call upon vast numbers of helicopters to support its ground troops, from relatively basic transport types such as the Mil Mi-8 'Hip' to highly capable combat types such as the Kamov Ka-52 'Hoodlum' attack helicopter. While at sea, variants of the Kamov Ka-27 'Helix' provide the most commonplace aerial assets of the Russian Navy.

This chapter includes the following aircraft:

- Kamov Ka-27
- Kamov Ka-29
- Kamov Ka-31
- Kamov Ka-52
- Kamov Ka-226
- Kazan Ansat
- Mil Mi-8
- Mil Mi-14
- Mil Mi-24 and Mi-35 family
- Mil Mi-26
- Mil Mi-28
- Mil Mi-38

Though other helicopter producers have come to the fore in recent years, the Russian helicopter fleet is still dominated by the products of the Mil design bureau. The most recent Mil combat design is the fearsome Mil-28N, optimized for destroying armour by night.

Kamov Ka-27 NATO reporting name: Helix

The Kamov Ka-27 is a ship or shore-based anti-submarine and search-and-rescue helicopter with a normal crew of three, and Kamov's trademark coaxial, contra-rotating rotor configuration.

Work on the design began at the Kamov design bureau in Lyubertsy in 1968 and was followed in 1972 by an official order for a Soviet Navy helicopter to supersede the Ka-25 Hormone. Originally known as the Ka-252, the new helicopter was required to be on station for two hours at a distance of 200km (124 miles) from the ship and the Navy required the new helicopter to have the same dimensions as the Ka-25, allowing it to operate from the same vessels.

Series production

The maiden flight of the first prototype took place on 8 August 1973, and during 1977 and 1978 five prototype and pre-series helicopters undertook state acceptance tests on board the aircraft carrier *Minsk*. State acceptance tests were completed by the end of 1978 and were followed by the launch of series production in July 1979. The new helicopter was officially accepted into Soviet Navy service as the Ka-27 in April 1981 for the anti-submarine warfare role.

The original Ka-27 (izdelye D2 or *izdeliye* 500), Helix-A, featured the Osminog automated search/attack system, consisting of an Initsiativa-2KM radar, VGS-3 Ros'-V dipping sonar, tactical display and digital computer. It could optionally be fitted with the APM-73V Bor towed magnetic anomaly in place of the sonar. The Ka-27 also features a radio sonobuoy system, separate to the Osminog system, with airborne receiver and RGB-NM-1 sonobuoys. A data transmission system connects the mother ship and a group of up to four helicopters. Weapons options were one torpedo, including APR-2 rocket-powered torpedo, or 6-8 250kg (550lb) depth charges, including the Zagon series guided depth charges, or one nuclear depth charge. Up to 36 sonobuoys can be carried.

State acceptance tests

The basic Ka-27 was followed by the Ka-27PS (*Poiskovo-Spasatelnyi*, search-rescue, izdelye 501), Helix-D, search and rescue version. With all

---

Kamov Ka-27

Weight (maximum take-off): 11,000kg (24,251lb)

Dimensions: Fuselage length 11.3m (37ft 1in), Rotor diameter 15.9m (52ft 2in), Height 5.4m (17ft 8in)

Powerplant: Two Klimov TV3-117KM turboshaft engines each rated at 1618kW (2200shp)

Maximum speed: 290km/h (180mph)

Range: 700km (435 miles)

Ceiling: 3500m (11,480ft)

Crew: 1-3 plus 1-3 specialists

Armament: Up to 2000kg (4409lb) of weapons and stores

Ka-27M

Part of the Baltic Fleet, '41 Yellow' is a modernised Ka-27M of the Russian Navy, based at Chernjahovsk air base in the Russian enclave of Kaliningrad during 2018.

Ka-27PS

'33 Yellow' is on the strength of the 318th Independent Composite Aviation Regiment at Kacha Airbase in Crimea and took part in large scale exercises in the Caucasus during 2016. Note the air-sea rescue hoist mounted outside the rear cabin door.

anti-submarine equipment removed (but retaining the Osminog-PS radar), the Ka-27PS carries two radio signal buoys below the tail boom, and a winch at the cabin door, as well as life-rafts and containers with clothes, food and medicine, etc. Also used as utility transport, the Ka-27PS was first flown on 8 August 1974 and completed its state acceptance tests between 1978 and 1979. An extended range variant, the Ka-27PSD (*Dalnyi*, long-range), featured increased fuel and strengthened undercarriage. An unusual variant of which few were constructed is the Ka-27Ye, used for radiological reconnaissance, with sensors to detect nuclear weapons on board ships.

Manufactured by the Kumertau Aircraft Production Enterprise, around 280 Ka-27s including export Ka-28 versions were built, but after the collapse of the Soviet Union only low-rate production for export was undertaken, finishing with a batch of

A Russian Navy Ka-27PS lands on the deck of USS *Mount Whitney* (LCC-20) as part of international partnership training during FRUKUS 2010.

Ka-28s for China completed in 2009 and 2010. Around 70 anti-submarine and 15 search-and-rescue versions are in Russian Navy service, and these are currently undergoing an upgrade programme to Ka-27M standard launched in 2015 as part of regular repair work for Ka-27s at Kumertau. The national armament programme called for 46 helicopters to be upgraded by 2018.

The Ka-27M retains the original rotor system, powerplant and airframe but adds an all-new mission system, the Bumerang, derived from the Novella system used in the Il-38N. Bumerang comprises a new radar, radio sonobuoy system and buoys, dipping sonar, fixed magnetic anomaly detector, electronic support measures and tactical display. Weapons include the APR-3 anti-submarine missile and new navigation and communication equipment are fitted.

Active service

Tests of the first upgraded Ka-27M were completed in 2011. These helicopters serve aboard the aircraft carrier *Admiral Kuznetsov* (up to 28), with one or two examples onboard missile cruisers, destroyers and frigates. Around 15 Ka-27s serve with the Russian Border Guard to patrol maritime boundaries, while the Russian Aerospace Forces has a handful for search and rescue.

A small number of (ex-Soviet) Ka-27s remain in use with the Ukrainian Navy as of 2022. Export-variant Ka-28s (*izdeliye* 330) were exported to China (17), India (14), Vietnam (10) and Yugoslavia (1). The Ka-28 was first ordered by India and possesses a different radio sonobuoy system with new sonobuoys, increased maximum take-off weight and increased maximum fuel load. The K-28M is a mid-life upgrade offered for export Ka-28s.

The first true combat role for a Russian Navy Ka-27 took place during the Syrian civil war. When *Admiral Kuznetsov* took part in Syrian operations between November 2016 and January 2017, the air group included Ka-27s for anti-submarine warfare and Ka-27PS for search and rescue/utility. Also embarked were Ka-29s and Ka-31Rs.

This 'Helix' was assigned to the Russian destroyer *Admiral Vinogradov* and was photographed from the US guided-missile cruiser USS *Vella Gulf* while conducting operations in the Gulf of Aden in 2009.

Kamov Ka-29 NATO reporting name: Helix-B

A ship-based assault helicopter able to transport assault troops from ship to shore, the Ka-29 is also capable of providing fire support for naval infantry.

The Ka-27 was used as a platform from which to develop a range of different specialist helicopters, including the Ka-29 for naval assault, the Ka-31 radar picket and Ka-32 civilian helicopter. All shared the same basic propulsion system and portions of the avionics, but introduced different cabins and mission systems. The Ka-29 features a wider, more angular forward fuselage containing a cabin with space for 12-16 troops, or four stretchers with eight passengers. Additionally, it is capable of carrying a load of up to 4000kg (8820lb) in a sling. The maiden flight of the Ka-29 prototype took place on 28 July 1976, trials were completed during 1979 and the helicopter was officially commissioned into Soviet Navy service in August 1987.

The Kumertau Aircraft Production Enterprise plant built 59 of the basic production version Ka-29 (Ka-252TB (*Transportno-Boyevoi*, combat-transport or *izdeliye* D2B, *izdeliye* 502), between 1984 and 1993. The primary role of the aircraft was to operate from the four Project 1174 Ivan Rogov-class assault ships, each of which could carry four helicopters. All these ships are now retired and most Ka-29s have been placed into storage. Survivors are used mainly for transport, although some have been transferred to the Federal Border Service. Ukrainian Navy Ka-29s are no longer believed to be operational but at least one example was transferred to Equatorial Guinea.

Ka-29TB

The portly dimensions of the Ka-29 are dictated by hangar size constraints aboard Russian Navy ships. '39 Yellow' took part in training operations during 2016 with the 859th Naval Aviation Training Center at Yeysk on the Black Sea.

The same aircraft as depicted in the profile above, '39 Yellow' was photographed in July 2021 over St Petersburg.

In fire support configuration, the main armament of the Ka-29 is the Shturm-V anti-tank missile system as also used on the Mi-24. This adds electro-optical sight and radio-command guidance antenna under the nose plus up to eight 9M114 (AT-6 Spiral) tube-launched anti-tank missiles. Unguided weapons options include up to four pods for 80mm (3.1in) unguided rockets, UPK-23-250 gun pods, bombs or incendiary tanks. An optional 2A42 30mm (1.2in) single-barrel cannon with 250 rounds can be fitted on the port weapons rack.

By contrast, in assault transport configuration the Shturm-V is removed, but a GShG-7.62 4-barrel Gatling-type 7.62mm (0.3in) machine gun and 1800 rounds of ammunition is fitted in the front of the cockpit on the starboard side.

TV/optical targeting

The Ka-29 VPNTsU (*Vozdushnyi Punkt Nabudeniya, Tseleukazaniya i Upravleniya*, airborne surveillance, target indication and control post) was used during the Chechen conflict to support Ka-50 combat helicopters on an experimental basis. The VPNTsU version added the TV/optical targeting system from the Ka-50 under the nose.

The Russian order for four Mistral-class amphibious assault ships from France led to plans to overhaul and upgrade some of the Ka-29 fleet, returning them to service in the more advanced Ka-29M standard. These would have had new avionics, including a glass cockpit, new radar and forward-looking infrared turret, but the Mistral order was cancelled following Russia's annexation of Crimea in 2014 and no Ka-29M has yet appeared. Ka-29s have seen operations as part of the aviation component carried by the carrier *Admiral Kuznetsov* deployed in Syrian waters during late 2016 and early 2017.

Kamov Ka-29

Weight (maximum take-off): 11,500kg (25,353lb)
Dimensions: Fuselage length 11.3m (37ft 1in), Rotor diameter 15.9m (52ft 2in), Height 5.4m (17ft 8in)
Powerplant: Two Klimov TV3-117V turboshaft engines, each rated at 1618kW (2200shp)
Maximum speed: 290km/h (180mph)
Range: 460km (286 miles)
Ceiling: 4300m (14,108ft)
Crew: 3
Armament: One mobile forward firing 7.62mm (0.3in) Glagolev-Shipunov-Gryazev GShG-7.62 four-barrelled machine gun, one optional 30mm (1.2in) Shipunov 2A42 autocannon fixed forward firing under fuselage; up to 2000kg (4409lb) of weapons and stores

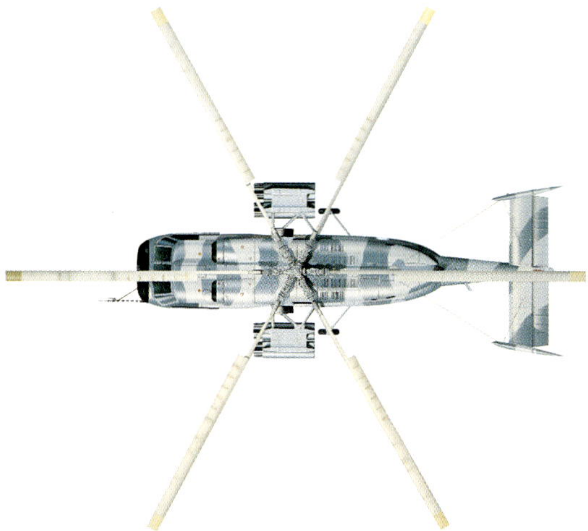

Ka-29TB
One of the pre-production development airframes, '25 Blue' featured different instrumentation for flight test operations and lacked the fuselage-mounted 30mm (1.2in) cannon fitted as standard to production machines.

Kamov Ka-31 NATO reporting name: Helix

The Ka-31 is a radar picket helicopter developed for the Soviet (later Russian) Navy, for detection of airborne and sea targets. Subsequently, a version of the Ka-31 was developed for battlefield surveillance.

In 1975, development began at Kamov in Lyubertsy of a shipborne radar picket helicopter based on the Ka-27 (then still known as Ka-252). Initial plans called for a battlefield surveillance version too, but both models were abandoned. However, the programme was resurrected in the mid-1980s with new mission systems as the naval Ka-252RLD and land-based Ka-252SV. The radar picket variant was intended as a stopgap until the carrier-based turboprop Yak-44 AEW aircraft was available but in the event the Yak-44 was itself abandoned on the collapse of the Soviet Union.

The prototype of the Ka-252RLD made its first flight on 25 November 1986 and tested the antenna and undercarriage, although it was not fitted with mission systems. Two fully equipped prototypes were completed in 1990 and 1991, and state acceptance tests were finally completed in 1996. The Ka-31 (*izdeliye* 03D2, *izdeliye* 503, also 531) is the standard production version for export. Primary sensor is the E-801 pulse-Doppler radar, stowed horizontally below the fuselage for cruise flight and lowered for operating. Reportedly, this radar can track up to 40 airborne or surface threats simultaneously, detecting a fighter-size target at up to 115km (70 miles). The Ka-31 features a modified fuselage with a widened nose and forward cabin, as seen on the Ka-29. An all-new landing gear is fitted with four legs that are raised hydraulically when radar is deployed.

Indian contract

The first order for the new helicopter came from India in August 1999, covering four Ka-31s to serve aboard the aircraft carrier *Viraat* and three Krivak III-class frigates. A contract for five further helicopters was signed in February 2001.

Kamov Ka-31

Weight (maximum take-off): 12,500kg (27,558lb)

Dimensions: Fuselage length 11.3m (37ft 1in), Rotor diameter 15.9m (52ft 2in), Height 5.64m (17ft 8in)

Powerplant: Klimov TV3-117VMAR turboshaft each rated at 1618kW (2200shp)

Maximum speed: 250km/h (155mph)

Range: 600km (373 miles)

Ceiling: 3500m (11,483ft)

Crew: 2

Armament: None

Ka-31

'32 blue' appeared at the MAKS airshow in August 2005. Seen here as it appears in flight with radar deployed and undercarriage raised, the Ka-31 presents a very distinctive appearance in operating mode.

Manufactured by Kumertau Aircraft Production Enterprise, the first series-production helicopter was flown on 16 May 2001 and all nine were delivered to India between 2003 and 2004.

Another Indian order followed in August 2009 covering five more Ka-31s that were delivered in 2011 and 2012. In September 2007 China ordered nine Ka-31s, delivered between 2010 and 2012.

An order for the Russian Navy was placed in November 2008 for the Ka-31R version. Likely features of the Ka-31R include software improvements and Russian-standard communications and friend or foe identification. After some delays, the first of two trial airframes was delivered to the naval aviation test centre at Yeysk in June 2012. A total of 12 Ka-31Rs were covered by the national armament programme up until 2020 and it was planned to issue these helicopters to the Northern and Pacific Fleets. However, it would appear that no further series production aircraft have been noted as of 2022.

The Ka-35 (*izdeliye* 23D2; Ka-252SV for *Sukhoputnye Voyska*, the Army) retains the Ka-31's airframe, powerplant, transmission and rotors and is intended as part of a battlefield reconnaissance and target acquisition system for the ground forces, conducting reconnaissance of ground targets and transmitting data to command posts and to serve as data relay between attack helicopters, unmanned air vehicles and ground posts. The first example of the Ka-35, which also adds a new self-defence system, was converted from the first Ka-31 prototype, while the second was built as new, but there is no confirmation of any further orders.

Ka-31

With radar closed up and undercarriage lowered, the Ka-31 appears little different to the Ka-29. '90 Red' was delivered to Yeysk Naval Aviation Evaluation Centre in June 2012 later taking part in the Air Parade on Navy Day 2018 in St Petersburg.

In June 2015, Ka-31 '91 Red' was found at the Kubinka air base during the annual 'Army Forum' display. From this angle, and folded up against the lower fuselage the radar is virtually invisible and there is little clue as to the helicopter's true function.

Kamov Ka-52 NATO reporting name: Hokum

A side-by-side two-seat reconnaissance/combat helicopter with the primary role of battlefield reconnaissance, the Ka-52 can also coordinate helicopter strike groups and attack ground and aerial targets.

Ka-52

Hokums are finished in overall dark grey or this somewhat arid camouflage scheme of sandy brown and olive drab over bright blue undersides. '76 Yellow' is part of the 18th Brigade of Army Aviation at Khabarovsk Tsentralnyi, a unit that began to receive the Ka-52 during 2014–15.

Ka-52

'22 Yellow' of the 319th Independent Helicopter Regiment (319 OVP) based at Chernigova. This unit was the first to operate the Mi-24 in 1971. 319 OVP currently consists of a squadron of Mi-8AMTSh transport helicopters, a squadron of Ka-52s and a handful of Mi-26s.

Designed by Kamov in Lyubertsy and manufactured by the Progress Company in Arsenyev, today's Ka-52 was developed from the single-seat Ka-50 (Hokum-A). The Soviet government had launched work on a new-generation combat helicopter in December 1976, which resulted in both the Mi-28 and Ka-50, and the first hovering flight by Kamov's prototype single-seat V-80 took place on 17 June 1982. The first series-production aircraft from the initial single-seat batch of 12 was flown in May 1991, now designated Ka-50 (*izdeliye* 805). Lacking fire-control radar, the Ka-50 utilized an electro-optical targeting system for its Vikhr missiles. Two Ka-50s were used for combat trials in the Chechen campaign in early 2001 and around six were used for evaluation work as of 2015, but are now apparently withdrawn. Work on the two-seat version, V-80Sh2 (*izdeliye* 800Sh2 or *izdeliye* 806), was launched in 1987, adding night and all-weather capability. The first prototype Ka-52 Hokum-B was

Kamov Ka-52
Weight (maximum take-off): 11,300kg (24,912lb)
Dimensions: Fuselage length 13.87m (45ft 6in), Rotor diameter 14.5m (47ft 6in), Height 5.05m (16ft 6in)
Powerplant: Two Klimov/St Petersburg VK-2500 turboshafts, each rated at 1790kW (2400shp)
Maximum speed: 300km/h (186mph)
Range: 460km (286 miles)
Ceiling: 5500m (18,045ft)
Crew: 2
Armament: One 30mm (1.2in) Shipunov 2A42 auto-cannon semi-rigidly mounted on starboard fuselage side; up to 2000kg (4409lb) of weapons and stores

Photographed in 2009, '061 Yellow' is caught with the undercarriage just beginning to retract. The tall co-axial rotors cancel out torque and remove the need for a vulnerable tail rotor, in the event of the rudder becoming inoperable the helicopter remains airworthy.

converted from a production Ka-50 and had its maiden flight on 25 June 1997, although the programme progressed very slowly, with Kamov initially looking for export customers in the absence of Russian orders. In late 2006 the Ka-52 programme was relaunched, with the Russian Ministry of Defence outlining new requirements. In 2007, contracts were placed for two test Ka-52s and a batch of 24 helicopters (*izdeliye* 826); the test helicopters were completed in 2008 with the first pre-series models completed in 2009. By May 2011, the first deliveries were made to an operational unit at Chernigovka.

Laser guidance

Unlike other helicopters in its class, the Ka-52 utilizes Kamov's trademark coaxial, contra-rotating rotor configuration and features a crew escape system using ejection seats. The BREO-52 mission system is based around the Myech-1U radar (exported as Arbalet-52) and GOES-451.24 electro-optical turret (thermal imaging camera, TV camera, laser rangefinder/designator, laser beam riding for missiles, plus laser spot tracker), and Vitebsk-52 self-defence suite. The latter includes radar, laser and ultraviolet warning sensors, as well as radar and infrared jammers, flares and radar decoys. Stores are carried on six underwing pylons, of which the outer two can only carry 9M39 Igla-V (SA-18 Grouse) air-to-air missiles (two per pylon). The primary weapon system is the Shturm-VU anti-tank guided missile system compatible with 9M120-1 Ataka-1 (AT-9 Spiral-2) and 9A4172K Vikhr-1 (AT-16 Scallion) missiles. Both utilize laser beam-riding guidance, while the 9M120 can also use radio-command guidance. Missiles are loaded in clusters of up to six, with a normal maximum load of 12 missiles.

Unguided weapons include up to four launchers for 80mm (3.1in) rockets and the aircraft mounts one fixed Shipunov 2A42 30mm (1.2in) cannon with up to 460 rounds on the starboard side of the fuselage.

In March 2011, the Russian Ministry of Defence ordered 143 Ka-52s to be delivered by 2018 and in February 2019 it was announced that Russia planned to buy 114 upgraded Ka-52M helicopters until 2027. The Ka-52M, first flown in prototype form on 10 August 2020, features modernized avionics, including more powerful computers and expanded capabilities in night-time operating conditions. Intended to carry new missiles and to be used in conjunction with unmanned aerial vehicles, future Ka-52Ms are expected to be fitted with a new V006 Rezets active electronically scanned array radar and OES-52 electro-optical targeting system. Future armament options include the Klevok (also known as Hermes-A), which is a further development of the Vikhr, Khrizanthema, which is an improved and enlarged Ataka, and the next-generation LMUR missile.

A maritime version, the Ka-52K (*Korabelnyi*, ship-borne; nicknamed *Katran*), was developed for the Mistral-class helicopter carriers ordered from France but later cancelled. In total 32 examples were reportedly ordered in 2014. Key features include shorter wings and folding rotor blades, emergency flotation gear, the option to carry Kh-35UV (AS-20 Kayak) anti-ship missiles and new navigation equipment and radar optimized for maritime targets.

Combat record

The Ka-52 has been extensively used in combat by Russian Aerospace Forces in support for the Syrian civil war. Ukrainian sources suggest that a Ka-52 (and a Su-30SM) took part in an attack on Ukrainian Navy warships in the Black Sea in November 2018, launching unguided rockets. Ka-52s have seen action during the 2022 Russian invasion of Ukraine with four helicopters lost within the first two weeks of the conflict.

The first and so far only export order was placed by Egypt, with 46 examples delivered between 2017 and 2019. Some of these are understood to be in Ka-52K configuration for shipborne operations.

Kamov Ka-226 NATO reporting name: Hoodlum

Developed as a general utility helicopter primarily for civilian use, the Ka-226 was adopted by the Russian military as a trainer.

Ka-226
On the original Ka-26 the engines were mounted in prominent nacelles on the end of the stub wings. On the newer Ka-226, the stub wings are a mounting point for the undercarriage and the engines have moved atop the fuselage.

A much-improved development of the Cold War era Ka-26, of which more than 800 were built, the Ka-226 was first flown in prototype form on 4 September 1997. The helicopter has a cockpit for a crew of two, with the option for a detachable module that can normally carry six passengers or a maximum of eight.

Series manufacture began in 2002 at the Strela plant in Orenburg although only the first 10 were built at this factory before production switched to the Kumertau Aircraft Production Enterprise (KumAPP). Most Ka-226 production has been for Russian state organisations, including the Ministry of the Interior, Ministry of Emergencies, and Federal Border Service, with others going to Gazprom. At first the basic Ka-226 was powered by two Rolls-Royce Allison 250-C20R/2 turboshafts, but from 2013 production aircraft were instead powered by two Turbomeca Arrius 2G1 turboshafts under the Ka-226T designation.

Dual controls

With the Russian Aerospace Forces, the Ka-226V is used at the training base in Sokolovyi. The Ka-226V (Ka-226.80) is the standard version for the Russian Ministry of Defence and features Allison engines, weather radar in the nose and dual controls. The first batch of 16 was ordered in 2011 and as many as 42 have now been delivered to the defence ministry.

Naval version

A naval derivative of the Ka-226T, the Ka-226TM was intended for the Russian Navy and Federal Border Service, although Navy plans have apparently been abandoned after the Mistral-class amphibious assault ships were cancelled.

The Ka-226 has been deployed with the Russian contingent in Syria. Some of these helicopters have apparently been donated to the Syrian military, with which they serve, albeit wearing Russian markings.

Kamov Ka-226
Weight (maximum take-off): 3400kg (7496lb)
Dimensions: Fuselage length 8.58 m (28ft 2in), Rotor diameter 13m (42ft 8in), Height 4.19m (13ft 9in)
Powerplant: Two Rolls-Royce Allison 250-C20R/2 turboshafts, each 465shp
Maximum speed: 220km/h (137mph)
Range: 600km (373 miles)
Ceiling: 5000m (16,404ft)
Crew: 1 or 2
Armament: None

Kazan Ansat

A lightweight general utility helicopter with two pilots seated side-by-side with dual controls, the Ansat-U (*Uchebnyi*, training) was developed for the Russian military under the *Pervokursnik* (first-year student) programme.

Designed and manufactured by Kazan Helicopters in Tatarstan, development of the Ansat was launched in 1993 as a new lightweight twin-engine helicopter. The maiden flight of the first prototype took place on 17 August 1999. Series production began in 2004 for civilian customers, but an early crash contributed to the fly-by-wire control system of the original version not being certified for passenger flights. As a result, in 2012 work began on Ansat-1M with conventional mechanical flight controls.

In September 2001, the Ansat-U was selected for the Russian armed forces, for basic and advanced training, to replace the elderly Mil Mi-2. State acceptance trials were completed in October 2009 and the first eight helicopters in the initial order were delivered during 2009 and 2010, initially to the Torzhok evaluation centre, later to the helicopter training base at Sokolovyi. In 2011, another batch of 32 Ansat-Us was ordered by the Russian Ministry of Defence.

Kazan Ansat

Pictured as it appeared during the Airshow to celebrate 100 years of the Russian Air force in 2012, '40 Yellow' is typical of the Ansat fleet which are currently finished in overall dark grey.

Reconnaissance-target indicator

Virtually all of the Ansats built thus far were of the Ansat-U, basic military training version, as described. However, during 2005, Kazan flight tested the Ansat-2RC (*Razvedchik-Celeukazatel*, reconnaissance-target indicator) featuring a reworked narrow fuselage with two tandem cockpits, forward-looking infrared turret, optional weapons and stores on four pylons, plus a fixed 12.7mm (0.5in) machine gun. Only one was built. In addition, a single example of the Ansat-LL (*Letayushchaya Laboratoriya,* flying test bed), was constructed. This helicopter was developed for testing missiles seekers on behalf of the Radar-MMS company.

Kazan Ansat-U
Weight (maximum take-off): 3300kg (7275lb)
Dimensions: Length 13.54m (44ft 5in), Rotor diameter 11.5m (37ft 9in), Height 3.56m (11ft 8in)
Powerplant: Two Pratt & Whitney Canada PW207K turboshaft engines with FADEC system, each rated at 477kW (640shp)
Maximum speed: 275km/h (171mph)
Range: 508km (316 miles)
Ceiling: 5700m (18,701ft)
Crew: 2
Armament: None

Mil Mi-8 NATO reporting name: Hip

The world's most produced helicopter, the Mil Mi-8 and its export variant, the Mil Mi-17, are still in production and thousands are in military and civilian service worldwide.

Designed by Mil in Moscow and manufactured by two factories (Kazan Helicopters and the Ulan-Ude Aviation Plant), the Mil Mi-8 is a combat transport helicopter that can fly missions including armed assault, transport, search and rescue, medical evacuation, minelaying and others. Further developed in specialized counterinsurgency, electronic warfare and airborne command post versions, the Mi-8 is normally flown by a crew of three: pilot, pilot/navigator, flight technician. Today, the major military versions for Russian service are the Mi-8MTV-5-1 from Kazan Helicopters and the Mi-8AMTSh from Ulan-Ude. They perform similar roles, but the Kazan version has a wider starboard door and normal-size port door. The Ulan-Ude version has a wider port door and normal-size starboard door. Both feature a rear ramp, absent from earlier Hips.

Troop-operated weapons

Standard accommodation is for 20 (Mi-8AMTSh) or 24 (Mi-8MTV-5-1) troops, optional extra seats in the centre of the cabin increase this to 34 (Mi-8AMTSh) or 36 (Mi-8MTV-5-1) troops. Armed versions have provision for six (Mi-8MT/MTV-2 and Mi-8AMTSh) or four (Mi-8MTV-3/ MTV-5) weapons and stores pylons. Typical armament includes four pods for 80mm (3.1in) unguided rockets, UPK-23-250 gun pods, sub-munitions dispensers, bombs and incendiary tanks. Provision is made for a flexible 7.62mm (0.3in) single-barrel machine gun in the nose on some versions, plus one at the cabin rear emergency door. Further troop-operated weapons can be fired through cabin windows.

Development of the Mi-8 began in January 1958 with plans for a new medium-weight turbine-powered transport helicopter to replace the piston-engined Mi-4. The first prototype of the single-engined V-8 (Hip-A) took flight on 24 June 1961. In the meantime, Mil had begun work on a larger version with two new TV2-117 turboshafts. This led to the V-8A (Hip-B) prototype, first flown on 2 August 1962. State acceptance tests were completed in November 1964, the first series-production Mi-8 was completed at Kazan in October 1965 and the helicopter was officially commissioned

Operating in the service of Kazakhstan, this Mil-8 'Hip-C' was carrying officials to observe a massed international parachute drop on Kazakh territory.

into military service as the Mi-8TV in 1968.

A second production line was launched at Ulan-Ude in 1970. Kazan built over 4000 Mi-8s with TV2-117 engines until 1980, then switched to the Mi-8MT with TV3-117 engines. Ulan-Ude built Mi-8s with TV2-117 engines until 1996, completing around 3700. As of 2000, Kazan had built 2787 helicopters with TV3-117 engines, while Ulan-Ude had built more than 100. The last of the first-generation Hips, with TV2-117 engines, were finally retired from Russian military service by 2009.

Multiple variations

The Mi-8 has been produced in a bewildering number of variants, some of the early TV2-117-powered Hips include the Mi-8T (V-8AT, *Transportnyi*), Hip-C, a standard civil and military transport with circular cabin windows; the Mi-8P (V-8AP, *Passazhirskiy*), Hip-C), passenger version with rectangular cabin windows; the Mi-8TV (*Vooruzhonnyi*, armed), Hip-C, with provision for four weapons pylons; the Mi-8MB *Bissektrissa*, medical evacuation variant was widely used in the Soviet Afghan War; the Mi-8SP and Mi-8SPA were space-crew rescue versions. A second version, confusingly also designated Mi-8TV, was a military assault transport with provision for six weapons pylons and a machine gun in the nose. Variants were also produced for artillery spotting, battlefield reconnaissance, minelaying radiation sampling, minelaying, electronic warfare and VIP transport.

A new generation of Hips was created in the early 1970s with new, more powerful TV3-117 engines, as used in the Mi-14 and Mi-24. The first of these was delivered in 1977 as the Mi-8MT (Hip-H), and widely exported as the Mi-17. Lessons of the Soviet war in Afghanistan led to the introduction of the TV3-117VM engine, optimized for 'hot and high' operations. This was installed in the Mi-8MTV-2, with provision for weapons pylons, and a nose machine gun, in production from 1988. Another round of improvements followed, also as a result of lessons learned from Afghanistan, with attention now focused on air assault capabilities. Earlier versions typically offloaded troops via one port fuselage-side door, since opening the rear clamshell doors cannot be done particularly easily or quickly. In the new Mi-8MTV-5 (exported as Mi-17V-5), the port fuselage-side door was larger and complemented by a new sliding door on the starboard side. At the same time, the rear door was replaced by a single-piece hydraulically operated flat ramp. The new-look nose contained a weather radar.

Low-cost versions

Series production began in 2000. Following the Mi-8MTV-5, Kazan Helicopters began to focus on lower-cost versions under the Mi-171A2 programme. These aircraft also

Mi-8T

A long-lived aircraft, thousands of Mi-8s have served dozens of countries since the mid-1960s. This Soviet example was built in 1975 and operated by Aeroflot until it was put into storage in 1993.

differ in being developed for civilian certification. The 'low-cost' versions derived from the Mi-8MTV-6 (Mi-17V-6) prototype, similar to the Mi-8MTV-5 but with VK-2500 turboshafts and new avionics, which were later used in Mi-17V-5 for the Indian Air Force. The Mi-171A2 (project designations Mi-8M or Mi-171M) is the baseline platform for future production with new Klimov VK-2500PS-03 engines. The composite main rotor is from the Mi-38 heavy transport helicopter and the X-type tail rotor is taken from the Mi-35M combat helicopter.

Mil Mi-8MTV-5-1
Weight (maximum take-off): 13,000kg (28,660lb)
Dimensions: Fuselage length 18.65m (61ft 2in), Rotor diameter 21.29m (69ft 10in), Height 5.54m (18ft 2in)
Powerplant: Two Klimov/St Petersburg TV3-117VM turboshaft engines, each rated at 1491kW (2000shp)
Maximum speed: 250km/h (155mph)
Range: 680-690km (423-429 miles)
Ceiling: 6000m (19,685ft)
Crew: 3
Armament: Up to 4000kg (8818lb) weapons and stores

The latest electronic warfare version for jamming enemy air defence radars is the Mi-8MTPR-1, using the Rychag system, based on Mi-8MTV-5 airframe. It was reportedly first delivered in 2013, with a contract for 18 more signed the same year. Electronic warfare versions built in smaller series (or one-offs) include: Mi-8MTI (Mi-13), Mi-8MTPI, Mi-8MTSh (-1, -2 and -3), Mi-8MTPSh, Mi-8MTD, Mi-8MTP, Mi-8MTPB, Mi-8MTR (-1 and -2), Mi-8MTU, Mi-8MTYa (-2), Mi-8MT-1S. All are codenamed Hip-K and based on the Mi-8MT. Roles include jamming air defence radars and radio datalink jamming, as well as communications intelligence and jamming.

Dedicated counterinsurgency (COIN) versions have been developed – the Mi-8MTKO (Mi-17N) reconnaissance helicopter – tested from 1999. It features an electro-optical turret (forward-looking infrared and laser rangefinder), new avionics suite with MFDs in the cockpit, satellite navigation receiver and night-vision goggles. This variant was used in Chechnya. The improved Mi-8MTKO-P featured a weather radar, helmet-mounted sight and image-processing system.

The Mi-8GU features two electro-optical turrets and the Mi-8MN, based on the Mi-8AMTSh, is part of Antares airborne reconnaissance-strike system. Fitted with Shturm-VK-8 anti-tank guided missile system with 9M114 and 9M120 missiles and with two electro-optical turrets. The Mi-8MNP-2 (*izdeliye* 825) has an improved Shturm-VK-8M missile system with separate surveillance and targeting turrets. Six examples were ordered by the Federal Security Service (FSB) in 2010.

Ministry of Emergency Situations

In 2011, the Russian Ministry of Defence placed new orders for Hips, including 140 Mi-8MTV-5-1 and 132 Mi-8AMTSh variants, and another 40 Mi-8AMTSh in 2013. Today, the Russian military has around 300 Mi-8 helicopters, plus approximately 40 electronic warfare versions.

These latter include Mi-8PPA, Mi-8SMV-PG and the new Mi-8MTPR-1 versions. Mi-8s are also widely used by other Russian state entities, including the Federal Security Service (FSB), Ministry of Interior (MVD), Ministry of Emergency Situations (MChS) and others.

Export customers

Around 100 countries operate Mi-8s, mainly for military use. Major recent customers include India, with an order for 80 Mi-17V-5 helicopters placed in 2008, and another 71 ordered in 2012.

Beginning in 2011, the United States ordered 63 Mi-17V-5 helicopters for the Afghan National Army Air Corps.

Mil Mi-14 NATO reporting name: Haze

This is a shore-based anti-submarine and search-and-rescue helicopter with amphibious capability and a normal crew of four: two pilots and a technician in the cockpit, plus a navigator/weapon systems operator in the cabin.

The programme was launched in April 1965 when the Soviet government ordered Mil to develop a shore-based anti-submarine helicopter able to destroy an enemy submarine at a depth of 400m (1312ft) and a speed of 32kts. The first prototype V-14 (*izdeliye* 140) completed its maiden flight on 1 August 1967 powered by TV2-117 turboshafts, as were the subsequent two aircraft. These were replaced by the more powerful TV3-117 engines, first flown in a Mi-14 in September 1969.

The first production helicopter from the Kazan factory was flown in January 1974 and in September 1975, and the first Mi-14s were delivered to an operational Soviet Navy unit at Donskoye, near Kaliningrad. The first variant in service was the Mi-14PL (*Protivo-Lodochnyi*, anti-submarine; *izdeliye* 140, V-14 or V-14E or export), Haze-A, the standard anti-submarine version with the Kalmar search/attack system comprising analogue mission computer, Initsiativa series radar, dipping sonar and datalink. Other equipment included a radio sonobuoy system and magnetic anomaly detector.

Hunter configuration

Weapons and stores are carried in a weapons bay below the cabin floor. In 'hunter' configuration, the Mi-14 carries up to 36 sonobuoys, the 'killer' variant is armed with one torpedo or eight 250kg (550lb) depth charges, or 12 50kg (110lb) depth charges, or a single nuclear depth charge, although some helicopters were upgraded

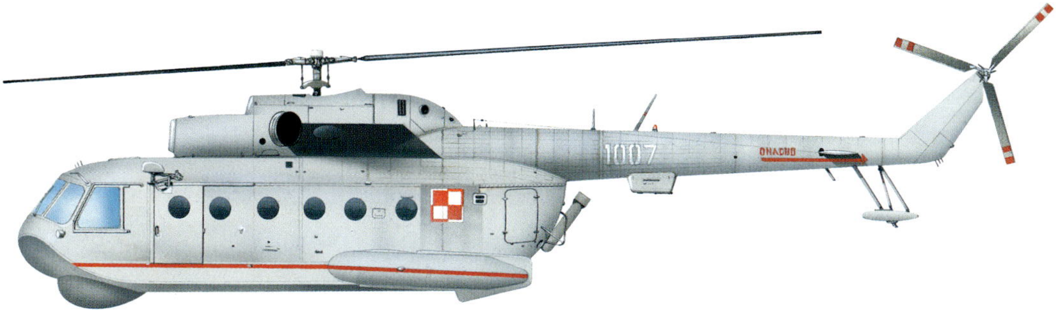

to carry APR-2 rocket-powered torpedoes. A separate stores bay is provided in the rear of the fuselage for day and night marker bombs, plus radar beacon buoys.

The ASW variant was followed by the Mi-14BT (*Buksir-Tralshchik*, tug-trawler), Haze-B, a mine countermeasures version and the Mi-14PS (Poiskovo-Spasatelnyi, search-rescue), Haze-C, search and rescue version featuring provision for 20 life-rafts, plus containers with clothes, food, medicines, etc. This variant was fitted with a larger cabin door and rescue hoist. The Mi-14PL/R (R is for *Ratowniczy*: rescue) is the Polish search-and-rescue version, upgraded from the Mi-14PL, which also features the larger cabin door and rescue hoist as well as a new weather radar in the nose, searchlights, GPS navigation and new communication devices.

Mi-14s were exported during the Soviet era to Bulgaria, Cuba, East Germany, Libya, Poland, Syria and Yugoslavia. Post-Soviet exports of second-hand helicopters went to Ethiopia, Pakistan and Yemen.

International exports

The official withdrawal from Russian Navy service took place in June 1996 and the helicopters are currently stored. However, both Poland (eight examples) and Ukraine (four in service as of 2020) still operate Mi-14s in Europe. Small numbers also serve with Syria and Pakistan. In June 2015, Russian Helicopters announced plans to return the Mi-14 to service after overhaul and partial upgrade. After this, production was to be relaunched. Funds were reportedly available for around 10 overhaul/upgrades. Later that year, the Russian Ministry of Defence said the 'new' Mi-14 would be used for future operations in the Arctic.

If production were to be relaunched, the next series of Mi-14 would incorporate new rotors, more powerful VK-2500 engines and all-new avionics. These plans seem to have been driven, at least in part, by Russia's annexation of Crimea, which put the repair facility at Sevastopol in Moscow's hands. This facility had handled maintenance of Ukrainian Mi-14s.

Polish upgrade

Separately, Poland has upgraded its Mi-14PL helicopters. Beginning in 2000, these received a computerized mission system including new sonobuoys, modernized dipping sonar and new magnetic anomaly detector, plus GPS and new identification friend-or-foe (IFF) equipment. Armament includes Italian A244/S anti-submarine torpedoes.

Mi-14PL

Poland is one of the current operators of the 'Haze'. This example operated from Darlowo Airbase during the 1990s.

Mil Mi-14PL

Weight (maximum take-off): 14,000kg (30,865lb)
Dimensions: Rotor diameter 21.29m (69ft 10in), Height 6.94m (22ft 9in)
Powerplant: Two Klimov TV3-117M turboshafts (M for Morskoi, seaborne, with special corrosion-resistant coating of some parts), each rated at 1454kW (1950shp)
Maximum speed: 240km/h (149mph)
Range: 1135km (705 miles)
Ceiling: 4000m (13,125ft)
Crew: 4
Armament: Up to 2000kg (4409lb) of weapons and stores

Mil Mi-24 & Mi-35 family NATO reporting name: Hind

The archetypal combat helicopter of the Soviet Union, the Hind remains in service with Russia and over 50 other nations. In modernized form it is still in production at Rostvertol in Rostov-on-Don.

Conceived in the late 1960s as a 'flying infantry fighting vehicle', able to carry a squad of soldiers while also providing its own suppressive fire, and with performance to ensure its protection, development of the Mi-24 was to be accelerated by using elements of the Mi-8 dynamic system combined with a new airframe and smaller tail rotor. Key features are a cabin seating eight troops, with access via horizontally split outward-opening doors.

Stub wings offload the main rotor in forward flight and boost speed, as well as carry weapons and stores. Two crew were initially accommodated side-by-side in a wide 'glasshouse' cockpit behind flat-plate glazing. The first prototype V-24 (*izdeliye* 240) was ordered by the Soviet government in May 1968 and first flown on 19 September 1969. This helicopter was designated A-10 when used to set a series of world records.

First production model

Series production was launched at Arsenyev, in the Soviet Far East, and the first production aircraft of the Mi-24A (*izdeliye* 245), Hind-A variant entered service with a Soviet Air Force training unit at Voronezh in 1970. With anhedral wings with wingtip pylons and featuring Falanga-P ATGMs with semi-automatic guidance, the Mi-24A was the first large-scale production version, with 240 built in 1972 to 1974.

The Mi-24U (*izdeliye* 244) Hind-C, was a training variant of Mi-24A and the Mi-24B (*izdeliye* 241) was similar to the Mi-24A, but added a four-barrel YakB-12.7 machine gun in the nose. Around 30 were built in the early 1970s. With the Mi-24D that entered production in 1973 (around 340 of which were built for the Soviets and around 300 for export, designated Mi-25), the helicopter was thoroughly redesigned, with a reshaped fuselage featuring separate tandem cockpits for the pilot and weapons system operator and a flight mechanic occupying a position in the passage between cockpit and cabin. This layout was retained on all subsequent Hinds.

The Mi-24DU, *izdeliye* 249, Hind-D, was a training variant of Mi-24D with no nose gun and the Mi-24V (*izdeliye* 242), Hind-E was similar to Mi-24D, but added the Shturm-V system with 9M114 ATGMs and later added TV3-117V for better high-altitude performance. This was the most

Mil Mi-24VP

The Mi-24VP replaced the standard under-nose turret-mounted 12.7mm (0.5in) machine gun of the Mi-24V with a 23mm (0.9in) Gryazev-Shipunov GSh-23 twin-barrel autocannon. With only 25 constructed, the Mi-24VP remains one of the rarest Hind variants in service. This Russian Navy example is based in Kaliningrad with the Baltic Fleet.

Mil Mi-35M

The Mil Mi-35M is to date the most advanced version of the 'Hind' series to have entered serial production, featuring updated avionics, targeting, and communication systems. As well as serving with the Russian Aerospace Forces, the Mi-35M has been exported; '03 Red' serves with the Air Defence Forces of Kazakhstan.

widespread version, with around 1400 built, and was exported as the Mi-35. Other Soviet era variants include the Mi-24P (*izdeliye* 243; P for *Pushechnyi*, cannon), Hind-F, known as Mi-35P for export, with a fixed twin-barrel 30mm (1.2in) cannon on the starboard side of the forward fuselage instead of the flexible nose machine gun. In total, 620 were built between 1981 and 1989.

The Mi-24VP (*izdeliye* 258), Hind-E, featured a nose-mounted flexible twin-barrel 23mm (0.9in) GSh-23 cannon instead of the YakB-12.7 machine gun. Unarmed variants include the Mi-24R (*izdeliye* 2462; R for *Razvedchik*, reconnaissance), Hind-G1, a nuclear, biological and chemical (NBC) reconnaissance/sampling version, and the Mi-24K (*izdeliye* 201; K for *Korrektirovshchik*, spotter), Hind-G2, utilized for artillery spotting and combat reconnaissance.

At its peak in 1990, the Soviet Union had 1600 combat Mi-24s and 300 more reconnaissance versions, and around 600 more were exported. Today, although the Ka-52 and Mi-28N are available, the Mi-35M is still prized by Russia for its troop-carrying and

Mil Mi-24VP Hind E

Weight (maximum take-off): 11,500kg (25,353lb)

Dimensions: Fuselage length 17.51m (57ft 5in), Rotor diameter 17.3m (56ft 9in), Height 5.47m (17ft 11in)

Powerplant: Two Klimov/St Petersburg TV3-117VM or TV3-117VMA turboshaft engines each rated at 1641kW (2200shp)

Maximum speed: 320km/h (199mph)

Range: 450km (280 miles)

Ceiling: 4500m (14,764ft)

Crew: 2 or 3

Armament: One 23mm (0.9in) Gryazev-Shipunov GSh-23 twin-barrel autocannon, flexibly mounted under nose; up to 1500kg (3307lb) of weapons and stores

casualty evacuation capability. It is also cheaper than these more modern attack rotorcraft and has been judged more reliable.

Around 2600 Mi-24s were built at Arsenyev until 1989. The factory in Rostov-on-Don began Mi-24 production in 1976, initially only for export, and by 1992 the Rostov plant had built around 700. Thereafter, low-rate production has continued at Rostov, most recently Mi-35Ms. The Mi-35M features a modified airframe with 'cropped' wings and fixed landing gear to reduce weight, plus new Klimov VK-2500 engines.

Other features include glass-fibre rotor blades, a new rotorhead with elastomer bearings and an X-type tail rotor, all taken from the Mi-28N. The Mi-35M was first ordered by Venezuela (10 examples delivered in 2006 to 2007), followed by Brazil (12), Azerbaijan (24) and Iraq (28). Finally, Russia placed two orders, the first for 22 helicopters in 2010, and another for 27 helicopters. These entered service with the evaluation centre at Torzhok. Unusually, the Mi-35M is also used as a VIP transport, the Mi-35MS (Salon, saloon) has large square windows and a side-opening cabin door, weather radar and a new communications suite. It carries no fire-control system or weapons.

In service

Today, Russian Aerospace Forces have around 45 of the new Mi-35Ms in use and approximately 80 Mi-24s, mainly Mi-24Ps, with some upgraded Mi-24PN versions. In the Mi-24P version, two weapons pylons are fitted under each stub wing, plus one below each wingtip endplate – the endplate pylons can only carry twin launchers for anti-tank guided missiles (ATGMs). Up to eight tube-launched 9M114 (AT-6 Spiral) ATGMs, including four on wingtip pylons and four on the two outer underwing pylons. The 9M32M Strela-2M (SA-7 Grail) anti-aircraft missiles can be carried. Unguided weapons on the four wing pylons can include bombs, sub-munition dispensers, gun pods, rocket pods and incendiary tanks. One twin-barrel 30mm (1.2in) GSh-2-30K cannon is fitted on the starboard side of the fuselage, with 250 rounds.

The Mi-24PN upgrade (N for *Nochnoy*, night-capable; *izdeliye* 543) was first flown in July 2000 with a new avionics suite including multifunction displays, digital map, satellite navigation and night-vision goggles. The ATGM system was modernized with 9K113N Shturm-VN version, with night-time capability and a nose-mounted infrared sensor. The shorter wing of the Mi-35M has no wingtip pylons. The 9M120

Despite being designed to combat US forces, this Mi-24 somewhat ironically now operates for the US Army Test and Evaluation Center, Threat Support Activity, in Las Vegas, Nevada, to provide simulated hostile threats to desert operations.

Mil Mi-24W

Weight (maximum take-off): 11,500kg (25,353lb)
Dimensions: Fuselage length 17.51m (57ft 5in), Rotor diameter 17.3m (56ft 9in), Height 5.47m (17ft 11in)
Powerplant: Two Klimov/St Petersburg TV3-117VM or TV3-117VMA turboshaft engines each rated at 1641kW (2200shp)
Maximum speed: 320km/h (199mph)
Range: 595km (370 miles)
Ceiling: 4500m (14,764ft)
Crew: 2 or 3
Armament: One 30mm (1.2in) Gryazev-Shipunov GSh-30-2 twin-barrel Gast autocannon flexibly mounted under nose; up to 2500kg (5510lb) of weapons and stores

Mil Mi-24W

The 'Hind' was one of the Soviet Union's most exported aircraft. This shark-mouthed Polish Mi-24H 'Hind-E' was operated by the 56th PSB at Inowroclaw during the Soviet era. Apart from the mouth, this 'Hind' has the standard Soviet Ground Forces aviation camouflage.

Ataka (AT-9 Spiral-2) ATGMs replace the 9M114s and are carried in two eight-round packs, for a maximum of 16 missiles. Up to four new 9M39 Igla-V (SA-18 Grouse) missiles are carried instead of Strelas. One twin-barrel 23mm (0.9in) GSh-23V cannon is mounted in an under-nose turret, with 450 rounds.

The Mi-24 saw extensive use by the Soviets in the Afghan War from 1979 to 1989, with experience there leading to improvements in construction, equipment and armament. After the demise of the Soviet Union, Mi-24s fought in conflicts on former Soviet territory including with Russia in Chechnya, in Dagestan, in the Armenian-Azerbaijani conflict, in Tajikistan and in the Russian war against Georgia in 2008. They also took part in the annexation of Crimea by Russia in 2014 and again in the Russian invasion of Ukraine in 2022, where the type has been used by both sides. Elsewhere, Mi-24s have taken part in dozens of local conflicts including the Iran–Iraq War, and conflicts in Angola, Ethiopia, Libya, Mozambique, Nicaragua and Syria. Algerian examples have been upgraded to the Super Hind Mk III by South Africa's Paramount Advanced Technologies with an observation/targeting turret in the nose, night-vision goggles, helmet-mounted sight, and new weapons include the Denel Vector 20mm (0.8in) cannon and Kentron Ingwe ATGMs.

Mil Mi-24K Hind-G2

CUTAWAY KEY

1 Remotely-controlled camera (Mi-24K) mounted beneath nose cannon barbette
2 Hinged lens cover
3 Low-speed precision airflow sensors
4 Air data sensor boom
5 IFF antenna
6 Armoured windscreen panel
7 Windscreen wiper
8 Gunsight
9 Pitot heads
10 Turret mechanism access doors
11 Ammunition feed mechanism
12 Four-barrelled 9-A-624 12.7-mm rotary machine-gun with 1,470 rounds
13 Gun turret; +20°traverse and +20°/- 60° elevation/ depression
14 Radar warning antenna
15 Retractable landing lamp
16 Boarding step
17 Kick-in steps
18 Oxygen bottle
19 Weapons Systems Officer (WSO) hatch
20 Collective pitch control column, flying controls duplicate in front cockpit
21 WSO's seat
22 Safety harness
23 Side console panel
24 WSO's hatch, open position
25 Cabin-mounted camera pallet (Mi- 24K)
26 Hinged lens aperture, replaces starboard cabin door
27 Film magazine
28 Cabin-mounted data-link console (Mi-24RCh)
29 Pilot's entry door, open position
30 Armoured windscreen panel
31 Windscreen wiper
32 Head-up-display
33 Instrument panel shroud
34 Cyclic pitch control column
35 Yaw control rudder pedals
36 Underfloor control linkages
37 Search light
38 Nosewheel leg door and indicator light
39 Levered suspension axle beam
40 Aft-retracting twin nosewheels
41 Conditioned air ducting (ammunition magazine on starboard side)
42 Nosewheel bay (semi-retracted housing)
43 Cockpit section armoured skin panelling
44 Cyclic pitch control lever
45 Fuel cocks
46 Circuit-breaker panel
47 Oxygen bottle
48 Rear view mirror
49 Pilot's seat
50 Engine air intake vortex-type dust/debris extractors
51 Debris ejection chute
52 Intake cowling

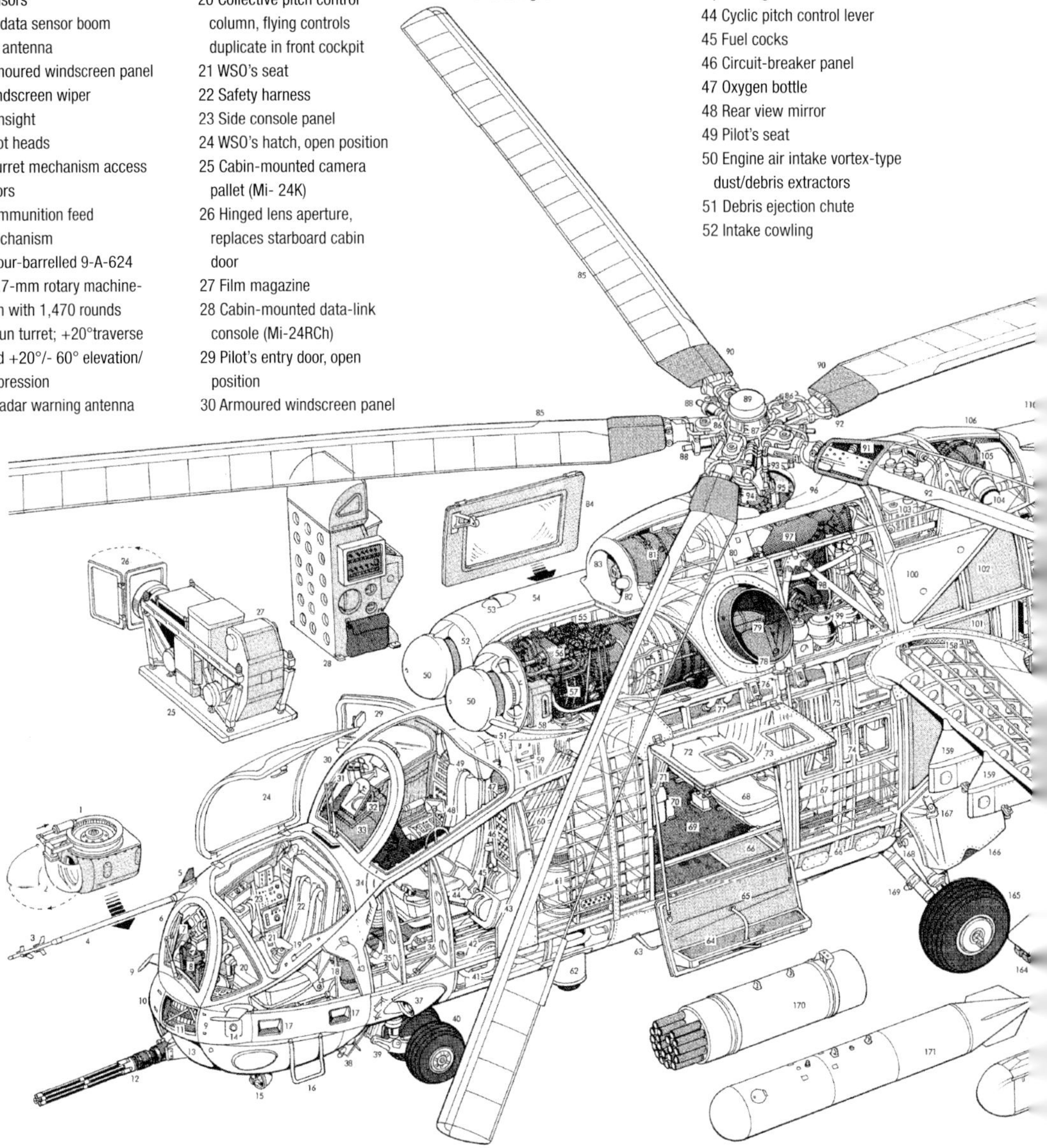

53 Generator cooling air intake
54 Starboard engine cowling/hinged work platform
55 Engine bay dividing firewall
56 Accessory equipment gearbox
57 Klimov TV3-117 turboshaft engine
58 Engine oil tank
59 Starboard side avionics equipment racks
60 Flight control rods
61 Cabin air filtration system equipment
62 Cabin air intake
63 Ventral aerial mast
64 Boarding step
80 Transmission oil
81 Oil cooler fan
82 Ice detector
83 Oil cooler air intake
84 Starboard upper door segment
85 Five-bladed main rotor
86 Blade root hinge joints
87 Titanium rotor head
88 Hydraulic drag dampers
89 Hydraulic reservoir
90 Blade root cuffs
91 Blade spar crack indicator (pressurized nitrogen-filled)
111 Aerial lead-in
112 UHF aerial mast
113 VHF aerial
114 Anti-collision light

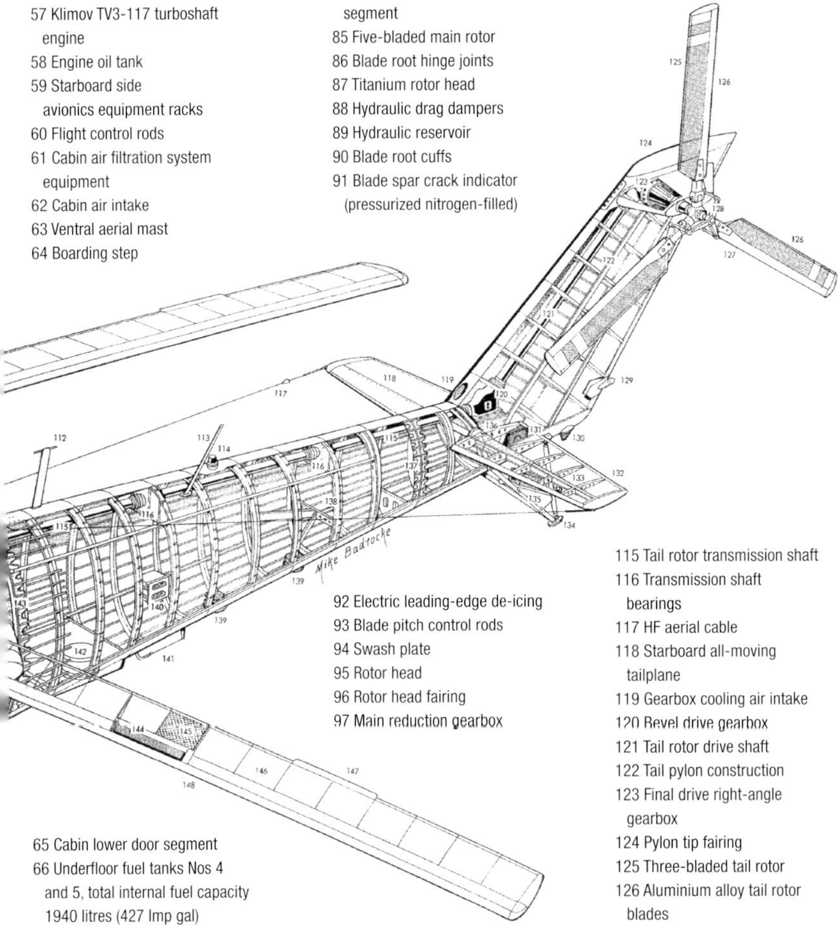

92 Electric leading-edge de-icing
93 Blade pitch control rods
94 Swash plate
95 Rotor head
96 Rotor head fairing
97 Main reduction gearbox

65 Cabin lower door segment
66 Underfloor fuel tanks Nos 4 and 5, total internal fuel capacity 1940 litres (427 Imp gal)
67 Cabin equipment consoles
68 Observer's seat
69 Main cabin floor panelling
70 Starboard side data-link console (Mi-24RCh)
71 Door interconnecting linkage
72 Tactical navigator's seat
73 Upper door segment
74 Cabin rear window panels
75 Cabin equipment racks
76 Winch mounting pads
77 Cabin air distribution ducting
78 Port exhaust duct
79 Engine/gearbox drive shaft
98 Gearbox mounting struts
99 Fire extinguisher
100 Fuel system equipment access
101 Main fuselage tank No. 3
102 Collector tanks nos 1 & 2
103 Rotor head control
104 APU exhaust
105 AI-9V auxiliary power unit
106 Hinged access panel
107 Air system vent
108 Auxiliary equipment
109 Generator
110 L-166V-11E microwave-pulse infrared jammer
115 Tail rotor transmission shaft
116 Transmission shaft bearings
117 HF aerial cable
118 Starboard all-moving tailplane
119 Gearbox cooling air intake
120 Bevel drive gearbox
121 Tail rotor drive shaft
122 Tail pylon construction
123 Final drive right-angle gearbox
124 Pylon tip fairing
125 Three-bladed tail rotor
126 Aluminium alloy tail rotor blades
127 Electric leading-edge de-icing
128 Blade pitch control mechanism
129 Tail navigation light
130 Lower IFF antenna
131 Flight recorder
132 Port all-moving tailplane
133 Tailplane rib construction
134 Tail bumper
135 Aft-facing camera mounting
136 Tailplane spar pivot mounting
137 Tail assembly joint frame
138 Tailcone frame and stringer construction
139 Radar altimeter antennas
140 Signal cartridge firing unit
141 Short-wave aerial cable
142 DISS-15D gyromagnetic compass units
143 Tailcone joint frame
144 Hollow D-section titanium blade spar
145 Honeycomb trailing-edge panels
146 Glass-fibre skin panelling
147 Fixed blade tab
148 Leading-edge anti-erosion sheath
149 Chaff-flare dispensers, port and starboard
150 Rear avionics equipment bay
151 Ventral access hatch
152 Ground power socket
153 Pneumatic system around connectors
154 Battery bay, both sides
155 Wing pylon tail fairings
156 Radar warning antenna
157 Stub wing rib and spar structure
158 Stub wing attachment joints
159 Port stores pylons
160 Port navigation light
161 Radar warning antenna
162 Stub wing ondplato pylon fairing
163 'Clutching-hand' ground sample collector
164 Hydraulically-operated scoops, three per side
165 Port mainwheel
166 Mainwheel leg door
167 Mainwheel leg aft pivot mounting
168 Shock absorber
169 Undercarriage indicator light
170 UV-32A-24 rocket launcher
171 PTB-450 110-Imp (500-litre) auxiliary fuel tank
172 Container store

Mil Mi-26 NATO reporting name: Halo

A very heavy transport helicopter, also capable of air assault missions, the Mi-26 remains the world's largest and heaviest helicopter to enter production.

Developed by the Mil design bureau in Moscow as a successor to the Mi-6, first flown in 1957, the Mi-26 adopted the same conventional layout but added a new powerplant and eight-blade main rotor. The first prototype made its maiden flight on 14 December 1977 and the state evaluation programme was completed in August 1980.

Civil variant

After three prototypes were built by Mil, series production was launched in 1980 by Rostvertol in Rostov-on-Don. The first production Mi-26 (*izdeliye* 90) was flown in October 1980 and the helicopter has been in series production ever since, although after the demise of the Soviet Union production was greatly reduced and is now mainly for export. In 1985 production of a civil variant began with the Mi-26T (*izdeliye* 209) civil transport helicopter. The Mi-26T dispensed with assault equipment, armour and military systems and added civil communication and navigation systems in their place. This was followed by the Mi-26TC (certified, or *izdeliye* 219), which was the same as the Mi-26T but with full civil certification. The Mi-26T2, an upgraded civil/military export version with new avionics, including navigation and communications, allowed the flight crew to be reduced from four to two. The prototype Mi-26T2 was completed in 2010 with the first order placed by Algeria, which ordered six helicopters in 2014.

The latest military version is the Mi-26T2V based on Mi-26T2 and boasts night-vision goggles compatibility, electro-optical turret and the Vitebsk self-defence suite (which includes missile launch warning sensors, infrared jammer and flare dispensers). By 2015, around 330 Mi-26s had been completed. The Russian Ministry of Defence placed a new order for Mi-26s in March 2011, with another batch of 18 that were delivered between 2011 and 2014.

Today the Russian Aerospace Forces operate around 40 Mi-26s, with many more in storage. Others serve with Federal Security Service, Federal National Guard Service and other

Mil Mi-26
Weight (maximum take-off): 56,000 kg (123,459lb)
Dimensions: Length 33.73m (110ft 8in), Rotor diameter 32m (105ft), Height 8.15m (26ft 9in)
Powerplant: Two Ivchenko-Progress/Zaporizhzhya D-136 turboshaft engines each rated at 7457kW (10,000shp)
Maximum speed: 295km/h (183mph)
Range: 800km (497 miles)
Ceiling: 4600m (15,090ft)
Crew: 5
Armament: None

state entities. In addition, Mi-26s were inherited by Belarus, Kazakhstan and Ukraine on the break-up of the Soviet Union, of which a handful of Kazakh examples are likely the only ones in use. The first export customer was the Indian Air Force in 1986, with other military deliveries following to Algeria, Cambodia, Democratic Republic of the Congo, Equatorial Guinea, Jordan, Mexico, North Korea and Venezuela. Some of these examples were supplied second-hand.

Operated by a standard flight crew of four: pilot, co-pilot, navigator and flight engineer, the Mi-26 features an unpressurized cabin capable of accommodating 68 fully equipped paratroopers or 82 troops. Alternatively, up to 61 litters plus three attendants may be carried. The aircraft can carry cargo loads of up to 20,000kg (44,092lb), including vehicles that may be loaded or unloaded via the rear ramp.

An alternative cargo capability is provided by the Mi-26TZ (*Toplivo-Zapravshchik*, fuel tanker), in production since 1996, which is configured to deliver 14,000 litres (3079 gallons) of fuel plus lubricants for the aircraft, tanks or other vehicles, for example at a forward arming and refuelling point (FARP).

Mi-26 in action

The Mi-26 was first used in combat by the Soviet Union in Afghanistan. It served in subsequent operations in Russian hands during post-Soviet conflicts Tajikistan, Armenia and Azerbaijan, Georgia, as well as Chechnya and Dagestan. The Chechen conflict saw one Mi-26 shot down by a man-portable surface-to-air missile in August 2002, killing 127 people in what was the biggest single loss of life in the post-Soviet Russian military.

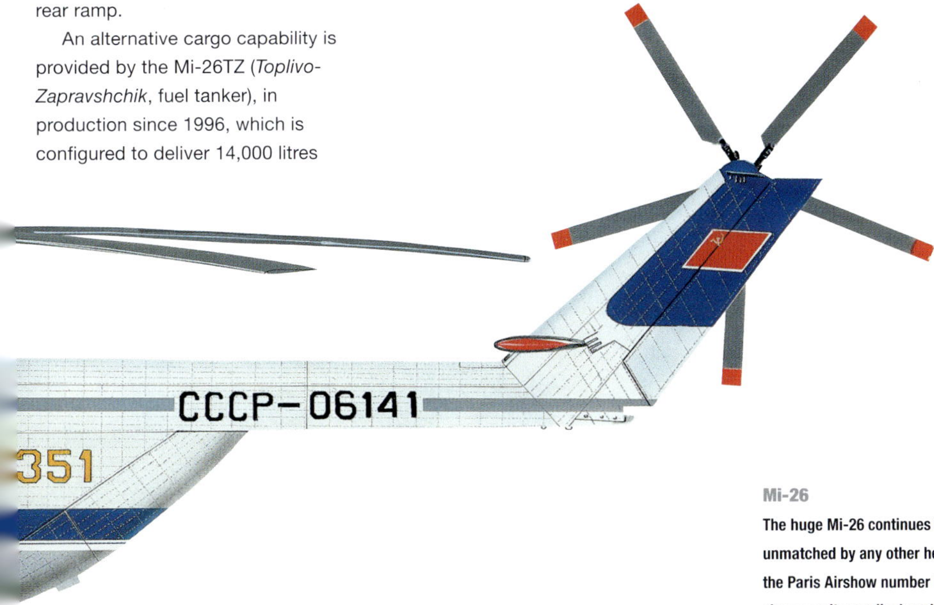

Mi-26

The huge Mi-26 continues to offer a capability unmatched by any other helicopter. Bearing the Paris Airshow number H-351, CCCP-06141 is shown as it was displayed in August 1981. It subsequently visited Farnborough Airshow in 1984.

Mil Mi-28 NATO reporting name: Havoc

Compared to the earlier Mi-24, the Mi-28 is optimized for the anti-tank mission, possessing improved manoeuvrability, performance and survivability.

In December 1976, the Soviet government ordered the development of a new-generation combat helicopter equivalent to the American AH-64 Apache. Preliminary approval was granted in June 1981 for Mil's design for an anti-tank and attack helicopter featuring a conventional layout for an aircraft of this type, with two crew-members seated in tandem, with the weapons system operator/navigator forward and the pilot at the rear. In contrast to the Mil Mi-24, which it superficially resembles, the Mi-28 has no secondary transport capability as the design dispensed with the cargo cabin in the fuselage. Two prototypes were completed by Mil in Moscow and designated Mi-28 (*izdeliye* 280), or Havoc-A to NATO, and the first of these made its maiden flight on 10 November 1982. Subsequently, the first of two prototypes of the redesigned Mi-28A version took to the air in January 1988 with new X-type tail rotor, redesigned engine exhausts, improved sensors and self-protection equipment. During the 1980s, Soviet authorities preferred the Ka-50 to the Mi-28 but Mil continued to fund Mi-28 development. In 1990, an agreement was signed with Iraq for export of the helicopter, as well as licensed production of the Mi-28L, although this was abandoned after the Iraqi invasion of Kuwait.

Night-capable version

In 1991, the Russian Ministry of Defence ordered Mil to develop a night-capable version, but funds were limited. This led to the Mi-28N (*izdeliye* 294) or Havoc-B, the base model for Russian forces, the first prototype of which flew in November 1996. Various problems led to the suspension of flight tests and work on new components. In 2002, the Mi-28N prototype returned to flight with new main transmission gear, followed by a second Mi-28N prototype in 2004, now with other new features including new main rotors, a new control plate and rotor head, and improved controls for engine and fuel system. Six pre-series helicopters were built by Rostvertol for further tests.

Production Mi-28N helicopters began to be delivered by Rostvertol to the then Russian Air Force in January 2008 when delivery was accepted of the first two production Mi-28Ns. These entered service with the evaluation and operational conversion centre in Torzhok. The Russian Ministry of

Mi-28N

On display at Kubinka in 2015, '52 Yellow' is finished in the standard tactical camouflage scheme worn by the Mi-28 fleet. In combat the Mi-28 has not proved invulnerable with at least one lost to a British-made Starstreak missile in the war in Ukraine.

Defence eventually placed orders amounting to 167 Mi-28Ns, 67 examples in 2005, 30 more in 2010 with a final order (to date) for 70 helicopters placed in 2011. Currently around 100 Mi-28Ns are in service with Russian Aerospace Forces. In addition, 40-60 examples of the Mi-28UB (*izdeliye* 298) are planned for Russian service, this being a version of Mi-28N with dual controls allowing for both training and combat missions. The first Mi-28UB flew for the first time in July 2013 and deliveries began during 2017.

Modernization programme

The latest development intended for Russian service is the Mi-28NM (*izdeliye* 296), a thorough modernization of the Mi-28N. The Mi-28NM features a new avionics suite, including new radar, electro-optical turret, laser missile homing system compatible with new versions of Ataka plus Khrizantema missiles and a new self-defence suite. Powered by VK-2500P engines with 'hot and high' capability, an initial test batch of eight much-modernized Mi-28NMs was ordered by Russia in December 2017 with plans to buy another 98. The first two series examples were flying by April 2019. A proposed intermediate version, known as the Mi-28D (*Dnevnoi*, daytime), would have used the Mi-28N airframe, powerplant and rotors/transmission and

Mi-28N

'04 White' features the overall grey finish applied to some Mi-28s, a finish that makes it look even more like the Boeing AH-64 Apache it superficially resembles. Compared to its US counterpart, the Mi-28 is slower but has a more powerful main gun and is better armoured.

Mil Mi-28N

Dimensions: Fuselage length 16.88 m (55ft 5in), Rotor diameter 17.2 m (56ft 5in), Height (without radar) 3.82m (12ft 7in)

Powerplant: Two Klimov/St Petersburg VK-2500 turboshafts, each rated at 1641kW (2200shp)

Maximum speed: 280km/h (174mph)

Range: 200km (120 miles)

Ceiling: 5000m (16400ft)

Crew: 2

Armament: One 30mm (1.2in) Shipunov 2A42 auto-cannon flexibly mounted under nose; up to 2350kg (5181lb) weapons and stores

---

Mi-28 'Havoc-A'

'022 Yellow' was the second prototype Mil-28, completed in 1983 and differed from later examples with its three-blade tail rotor. Three further prototypes were built to Mi-28A standard before development stalled in the early 1990s.

dispensed with night sensors but has not entered production.

The first export order for the Mi-28 came from Iraq, which ordered 15 Mi-28NE helicopters (without radar), receiving their first deliveries in 2014. The second export customer was Algeria, which ordered 42 Mi-28NE helicopters in a special variant with dual controls (similar to the Mi-28UB) and radar.

Mi-28s carry their stores on four underwing pylons. The standard anti-tank configuration comprises eight (or a maximum of 16) 9M120 Ataka-VN (AT-9 Spiral-2) anti-tank guided missiles carried in eight-tube clusters. Other weapon options include rocket pods, incendiary tanks, sub-munition dispensers, mine-laying containers and gun pods.

In addition 9M39 Igla-V (SA-18 Grouse) anti-aircraft missiles can be used against enemy helicopters. One single-barrel 30mm (1.2in) 2A42-2 cannon is mounted in an under-nose turret with 250 rounds.

Combat record

The Havoc saw its first combat with Iraq, during operations against so-called Islamic State insurgents. The Mi-28N has subsequently been used in combat by Russian Aerospace Forces during its intervention in the Syrian civil war, beginning with a first deployment in 2016, and again in the 2022 invasion of Ukraine.

'731 Yellow' flies past in unarmed condition. The contrast between the weathering of the rear and forward fuselage is dramatic, the area behind the engine being subject to the sooty exhaust from the aircraft's engines.

Mil Mi-38

A military version of the primarily civilian Mi-38 is expected to fulfil the requirement for the Perspektivnyi Sredniy Transportno-Desantnyi Vertolet (PSTDV), or future medium transport and assault helicopter. The programme also has the codename *Koptilshchik* (stoker).

Mi-38

Built in 2010, RA-38012 is the second prototype Mi-38 constructed by Kazan Helicopters. It is shown in the livery in which it appeared at the 2011 MAKS airshow.

Plans to create a new medium-sized helicopter, the Mi-8M, to supersede the Mi-8/17 were authorized by the Soviet government in July 1981 and work at Mil on the Mi-38 (*izdeliye* 260) programme then began in 1983. Various delays meant that the first prototype of the Mi-38, the civil Mi-38-1 model powered by two Pratt & Whitney Canada XPW127/5 turboshafts, only made its maiden flight on 22 December 2003. Progress continued to be sluggish due to lack of funding.

Over the following seven years, the helicopter made only 84 test flights. Nonetheless, a second prototype began flight tests in October 2010. A switch to the domestic Klimov TV7-117V engine was made by the time the third prototype, Mi-38-2 (*izdeliye* 2602), representing the Russian military configuration made its first flight in November 2013.

Cargo loads

The first customer was the Russian Ministry of Defence, which ordered two Mi-38T transport helicopters in July 2017, and both Mi-38Ts were handed over to the military in 2019. Plans exist to deliver two Mi-38s to the Special Air Group Rossiya that transports the state's highest authorities and Kazan Helicopters says it has more than 15 firm orders for Mi-38 from Russian state organizations.

The Mi-38 can accommodate up to 40 troops or 30 passengers. Cargo loads up to 5000kg (11,023lb) may be carried over a range of 420km (2601 miles), or 4000kg (8818lb) over 700km (435 miles). Up to 5000kg (11,023lb) can be carried externally.

Variants

Planned variants include the special forces Mi-383 that retains the powerplant, gearboxes and rotors, but features a new stealthy airframe and the Mi-38-7 (*izdielije* 2607), airborne command post, three of which have reportedly been ordered.

Mil Mi-38-2

Weight (maximum take-off): 16,200kg (35,715lb)
Dimensions: Fuselage length 19.95 m (65ft 4in), Rotor diameter 21.1m (69ft 3in), Height 5.2m (17ft 1in)
Powerplant: Two Klimov TV7-117VM turboshaft engines, each rated at 2087kW (2800shp)
Maximum speed: 320km/h (199mph)
Range: 920km (572 miles)
Ceiling: 6000m (19,685ft)
Crew: 2
Armament: None

TRANSPORTS

Russia operates the world's largest military transport aircraft, the Antanov An-124, but the workhorse of its military transport fleet remains the rugged Ilyushin Il-76. In the Beriev Be-200 amphibian, Russia possesses a transport aircraft with unique capabilities and the Ilyushin Il-112 project, currently in its development phase, should deliver a useful domestically-produced light transport.

This chapter includes the following aircraft:

- Antonov An-124
- Antonov An-140
- Antonov An-148
- Beriev Be-200
- Ilyushin Il-76 and Il-78
- Ilyushin Il-112

Currently the world's only jet-powered flying boat design in operational service, the Beriev Be-200 has proved particularly useful as a firefighting aircraft. This Be-200 was photographed taxiing out of the water onto a slipway at Gelendzhik, Russia in 2010.

Antonov An-124 NATO reporting name: Condor

The world's heaviest military transport aircraft, the An-124 is a long-range military freighter with a double-deck fuselage loaded through an upward-hinging fuselage nose and rear ramp.

The Soviet government ordered Antonov in Kyiv, Ukraine, to develop a new strategic military transport in February 1972, initially known as *izdeliye* 200 (since it was twice as heavy as the An-22, *izdeliye* 100). After one full-scale mock-up was produced, the programme was revised, calling for a more advanced aircraft with a related resolution of January 1977. This covered the *izdeliye* 400 that featured a new supercritical wing and fly-by-wire controls.

The first prototype made its maiden flight on 26 December 1982. The existence of An-124 was officially disclosed in May 1985, ahead of the aircraft setting a series of 21 world records. Production was launched in Kyiv where the first series production aircraft flew in December 1984. Meanwhile, the Aviastar-SP facility was created for An-124 production in Ulyanovsk, and the first aircraft from this source flew in October 1985.

Wings were 33 An-124s produced in Tashkent, Uzbekistan, and transported to Kyiv or Ulyanovsk on the fuselage of an adapted An-22. Two prototypes (one of which was static) and 54 series aircraft were completed (18 in Kyiv, 36 in Ulyanovsk).

Aircraft specifications

The Basic An-124-100 version carries up to 120,000kg (264,555lb), including containers, main battle tanks, heavy guns, missile systems, etc. Cargo can be paradropped, although paratrooper operations are constrained by the limited pressurization in the hold. The An-124 is standardly operated by a flight crew of six: pilot, co-pilot, navigator, two flight engineers and communication officer. In addition, two loadmasters attend to cargo handling. The upper deck of the fuselage includes two compartments for alternate crew and cargo attendants. The lower deck contains the cargo hold, accessed via a visor-type upward-hinging fuselage nose and rear ramp. The initial version was followed by the An-124-100 civilian version, certified for reduced weights, the An-124-100M, with crew reduced to four thanks to new avionics. The An-124-100M-150 is as the -100M but with maximum payload increased to 152

Antonov An-124
Weight (maximum take-off): 392,000kg (864,210lb)
Dimensions: Length 69.10m (226ft 8in), Wingspan 73.30m (240ft 6in), Height 21.08m (69ft 2in)
Powerplant: Four Ivchenko Progress D-18T series 3 turbofans, each rated at 229.8kN (51,654lbf)
Maximum speed: 850km/h (528mph)
Range: 14,200km (8823 miles)
Ceiling: 11,600m (38,058ft)
Crew: 6
Payload: 120,000kg (264,554lb)

An-124-100

A truly enormous aircraft, for several years Russia struggled to keep many of its 26-strong An-124 fleet operational. RF-82041 had languished on the ground since 1993 until it was returned to flying status in November 2014.

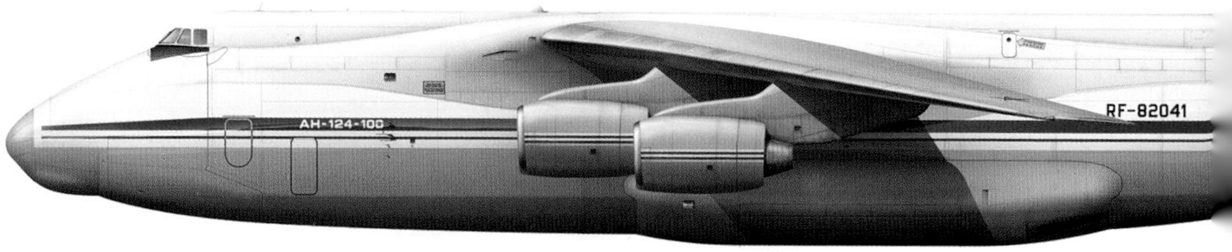

One of the Volga-Dnepr fleet of An-124s parked at Moffett Field in California, USA. US Forces contracted Volga-Dnepr to transport cargo with the An-124 to Afghanistan in April 2007 due to their own strategic transport fleet aircraft being fully engaged.

tonnes (150 tons), a single example of which is operated by Antonov, whereas the An-124-100-150 features increased payload but with the original avionics and six crew.

The first An-124 for the Soviet military was delivered in February 1987 to Seshcha and the aircraft was officially commissioned into service in March 1991. Today, Russian Aerospace Forces has around 10 airworthy An-124s from a total fleet of 26 aircraft. Non-flyable aircraft are being successively returned to service in upgraded An-124-100 form. The first aircraft to undergo this overhaul was returned to service in September 2010 after spending more than 10 years grounded.

Colossus programme

The largest civil operator is the Russian Volga-Dnepr company, which provides services for many governmental customers. Since 2010, Russia has had plans to resume production of the An-124 under the *Koloss* (Colossus) programme led by Ilyushin. This includes upgrades of Russian aircraft to An-124-100M (VTA) standard with new avionics and a glass cockpit and then production of new civil An-124-300 and military An-124-300V versions. These plans have been stalled since 2014 by conflict between Russia and Ukraine and economic crisis in Russia. The main problem is the current D-18T engine being manufactured only by Motor Sich in Ukraine. Preliminary work has been launched on Russian alternatives, starting with the Kuznetsov NK-23D turbofan based on the core of the NK 32 series 02 engine from the Tu-160, then also the Aviadvigatel PD-24.

The An-225 (*izdeliye* 402) Cossack, nicknamed *Mriya* (Dream), was a spectacular development of the An-124 originally designed to transport spacecraft components, including the Buran orbiter and Energiya rocket. The aircraft featured a lengthened fuselage, extended wing centre section, two more engines, added pairs of wheels and redesigned control surfaces. First flown 21 December 1988, it was the world's heaviest aircraft. The single example, used by Antonov since 2001, was destroyed during the Russian invasion of Ukraine in 2022.

Antonov An-140

Developed as a regional commercial turboprop, the An-140 is a 52-passenger airliner used by the Russian Ministry of Defence as a staff transport.

An-140-100

Built in 2014, this Russian Air Force An-140 was on display at the 2017 ARMY event at Kubinka, Russia. The An-140 programme has come under difficulty due to the aircraft's Ukrainian origins.

A product of the Antonov Design Bureau in Kyiv, Ukraine, the An-140 was originally schemed as a replacement for the An-24 short-haul regional airliner with improved fuel efficiency, longer range, higher speed and improved comfort. The first of two prototypes made by Antonov was flown on 17 September 1997.

Series production was launched in three locations, and production for Russian service is undertaken by Aviacor in Samara. Additional production takes place in Kharkiv, Ukraine and in Shahin Shahr, Iran. The first of 11 Kharkiv production aircraft was flown in October 1999, followed by the first of nine Iranian aircraft in February 2001 and the first of 12 from Samara in May 2005.

Staff transport

Between 2009 and 2013, the Russian Ministry of Defence ordered 14 An-140-100s from Samara (10 for the air force, four for the navy) as small staff transports. but the programme hit difficulties when conflict broke out between Russia and Ukraine in 2014. Production in Samara was then wound down and the last Russian-built aircraft was completed in 2016. Russian Aerospace Force aircraft are based in Chkalovsky and Yekaterinburg; Russian Navy aircraft at Knevichi and Yeysk.

There are currently three variants of the An-140: the initial version, designed to deliver 52 passengers 2200km (1367 miles), the An-140-100 is the standard production version with a longer wing and increased weight to deliver 52 passengers to 2400km (1491 miles), and the IrAn-140 Faraz version assembled in Iran. Several planned An-140 variants have been proposed, including the stretched An-140-200, the An-140-330T military transport with larger fuselage and Ivchenko AI-30 turboprops, the An-140C freighter and the An-140T (An-142), a military transport with rear ramp. The An-140MP is a proposed maritime patrol aircraft.

Antonov An-140

Weight (maximum take-off): 21,500kg (47,399lb)
Dimensions: Length 22.61m (74ft 2in), Wingspan 25.51m (83ft 8in), Height 8.23m (27ft)
Powerplant: Two 1864kW (2500ehp) Motor Sich TV3-117VMA-SBM1 turboprops
Maximum speed: 540 km/h (336mph)
Range: 3050km (1895 miles)
Ceiling: 7200m (23,622ft)
Crew: 2
Payload: 52 passengers or up to 6000kg (13,228lb) of cargo

Antonov An-148

The An-148 was developed as a commercial regional jet and used by Russian Ministry of Defence as an executive transport.

The An-148 programme was launched in 2001 to develop a successor to the Tu-134 regional jet, with the first flight taking place on 17 December 2004 in Kyiv. Series production has been undertaken by both Antonov in Kyiv, Ukraine, and Voronezh Aircraft Production Association (VASO) in Voronezh, Russia. The factories cooperated in production with final assembly lines at both facilities. Initially planned as a commercial aircraft, the An-148 has mainly been ordered by Russian state operators. The initial variants were the An-148-100, the basic airliner, which was followed by the An-148-100A and 100B that both featured the same passenger capacity but featured a higher maximum weight allowing for greater range capability.

State operators

In May 2013, The Russian Ministry of Defence ordered 15 An-148-100E aircraft from VASO for delivery between 2013 and 2017. The An-148-100E has a maximum weight of 43,700kg (96,342lb), allowing it to carry 75 passengers 4400km (2734 miles). The first aircraft built to this order was handed over to the Russian Ministry of Defence in December 2013. An-148s of the Russian Aerospace Forces are based at Chkalovsky, Levashovo and Rostov-on-Don.

Other Russian state operators include the Federal Security Service (FSB), and Special Flight Detachment Russia (SLO Rossiya). The Russian Ministry of Emergency Situations (Emercom) operates five An-148-100EM convertible versions: two passenger versions and three combined passenger-medical evacuation versions with two to six medical modules.

Production terminated after the breakdown in Russian-Ukrainian relations after 2014, with aircraft left incomplete in Voronezh received undercarriage from Russian production (previously the landing gear was delivered from Ukraine). Further development resulted in the An-158, a stretched 99-passenger version, in production in Ukraine only, and the An-178 military transport made by Antonov without Russian participation, which had its first flight in May 2015. Proposed An-148 variants include a 7000km (4349 mile) range executive jet and a maritime patrol aircraft.

Antonov An-148
Weight (maximum take-off): 43,700kg (96,342lb)
Dimensions: Length 29.13m (95ft 7in), Wingspan 28.91m (94ft 10in), Height 8.19m (26ft 10in)
Powerplant: Two Ivchenko D-436-148 turbofans, each rated at 66.97kN (15,050lbf)
Maximum speed: 870km/h (541mph)
Range: 4400km (2734 miles)
Ceiling: 12,200m (40,026ft)
Crew: 2
Payload: Up to 85 passengers or 9000kg (20,000lb) cargo

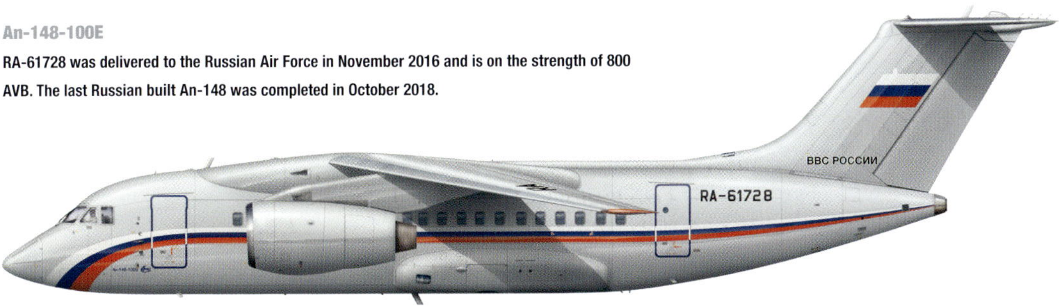

An-148-100E
RA-61728 was delivered to the Russian Air Force in November 2016 and is on the strength of 800 AVB. The last Russian built An-148 was completed in October 2018.

Beriev Be-200

A multi-purpose turbofan-powered amphibian, initially developed for search and rescue, the Be-200 is probably best known today as a civilian firefighting aircraft, although further variants are planned.

The Be-200 was developed by Beriev at Taganrog as a smaller version (roughly half the size) of the Beriev A-40 amphibian (NATO reporting name Mermaid) developed for the Soviet Navy in the 1980s. Originally known as the A-200 but later designated the Be-200, it performed its maiden flight (from land) on 24 September 1998 in Irkutsk, with the first take off from water following on 10 September 1999 in Taganrog.

Search and rescue

The Irkutsk aircraft plant made four Be-200 test aircraft and seven production Be-200ChS (*Chrezvychainykh Situatsiy*, emergency situations; *izdeliye* A201), a combined firefighting and search-and-rescue (SAR) aircraft. SAR equipment includes a weather radar in the nose, electro-optical turret under the port wing, a display at the observer's station in the cabin and a searchlight. Six Be-200ChS were delivered to Russia's *Ministerstvo Cherezvychainykh Situatsiy*, MChS, Ministry of Emergency Situations, between 2003 and 2011 and a single example was delivered to Azerbaijan in 2008.

During May 2011, the Russian MChS ordered a further six Be-200ChS aircraft to be built at Beriev's facility in Taganrog and scheduled to be delivered in 2013 and 2014, but the programme was considerably delayed due to Beriev's work on other programmes. The sixth and final Be-200ChS from the second MChS order was handed over in 2019, although one of the 12 aircraft was lost in a fatal accident while firefighting in Turkey in August 2021. In May 2013, the Russian Ministry of Defence signed a contract for six further Be-200s. Of these, two were planned as standard Be-200ChS and four in a new Be-200PS (*Poiskovo-Spasatelnyi*, search and rescue) version without the firefighting capability. Again, significant delays were incurred and the order was apparently later reduced to three Be-200ChS aircraft. The first Be-200ChS for the Russian Navy completed its maiden flight in February 2020. The Russian Navy has said it planned to acquire the anti-submarine warfare Be-200P version but this has not so far led to an order.

Firefighting

With a flight crew of two pilots in basic versions, plus optional observer and flight technician, the Be-200 can carry a payload of up to 5000kg (11,023lb) in the cabin. For SAR operations, the aircraft is capable of carrying up to 57 survivors/passengers in seats or up

Beriev Be-200ChS
Weight (maximum take-off): 41,000kg (90,390lb)
Dimensions: Length 32.05m (105ft 2in), Wingspan 32.78m (107ft 7in), Height 8.9m (29ft 3in)
Powerplant: Two Ivchenko Progress D-436TP turbofans, each rated at 73.55kN (16,535lbf)
Maximum speed: 700km/h (430mph)
Range: 3150km (1957 miles)
Ceiling: 8100m (26,575ft)
Crew: 2
Payload: Up to 57 passengers or 5000kg (11023lb) cargo or 12,000kg (26,455lb) water or retardant

Be-200ES

Azerbaijan became the first foreign customer for the Be-200 by purchasing a single example of the multi-role Be-200ES in April 2008. FHN-10201 is operated by the Azerbaijan Ministry of Emergency Situations as a fire-fighting, cargo or 43-seat passenger aircraft.

to 30 stretcher patients. Additionally, four droppable rescue capsules may be carried under the wing (containing a dinghy or life raft, plus medical, radio and other equipment). As a firefighter, up to 12,000kg (26,455lb) of water is contained in eight tanks under the cabin floor. Water can be dropped in a single salvo within one second, via doors in the hull.

Planned variants include the Be-210, a project for a passenger amphibian able to carry 72 passengers to 1850km (1149 miles). The Be-220 (or Be-220MP, *Morskoy Patrulnyi*, maritime patrol), is a planned anti-submarine and surveillance aircraft with an advanced mission system including sonobuoys, magnetic anomaly detector and electro-optical and electronic intelligence subsystems. Weapon options for this variant include depth charges and torpedoes carried in an internal bay and on underwing hardpoints. The Be-250 is a planned airborne early warning and target indication aircraft based on the Be-200, and the Be-300 is a planned shore-based derivative of the Be-200 amphibian with a new fuselage without the amphibious hull shape.

Be-200

'20 Yellow' is the first Be-200 built for Russian Navy use and delivered in February 2020. Named Aleksander Mamkin, the aircraft was lost with all eight on board on 14 August 2021 when it crashed whilst fighting fires in Turkey.

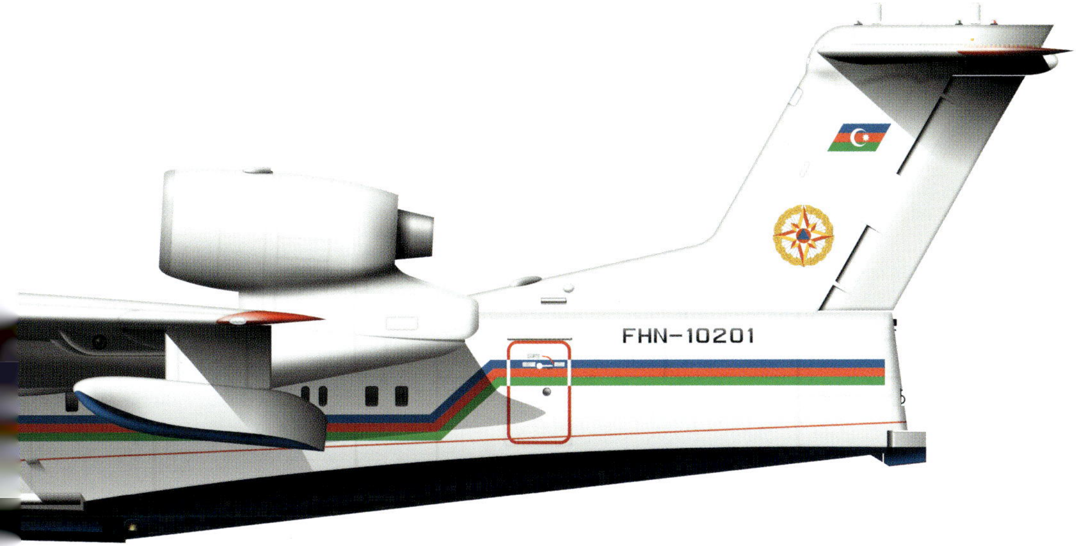

Ilyushin Il-76 and Il-78

NATO reporting names: Candid (Il-76) and Midas (Il-78)

The Il-76 is medium/long-range military transport, later converted for civil use, with specialized variants for medical evacuation, firefighting, astronaut training, electronic countermeasures and other roles.

In 1966, Ilyushin was tasked to design the Il-76 (*izdeliye* 76) as a transport for airborne assault, able to deliver a 33,000kg (72,752lb) load over a distance of 5000km (3106 miles), including operations from unprepared runways. The first prototype made its maiden flight on 25 March 1971 and the first series production aircraft was flown in May 1973, with the first aircraft delivered to the Soviet Air Force in June 1974. The aircraft was operated by a crew of six: two pilots, engineer, communications operator and loadmaster on upper deck, navigator on lower deck and an optional rear gunner. After around 80 aircraft were built, production switched in 1978 to the improved Il-76M, with a reinforced structure, more fuel capacity, widened rear fuselage and increased maximum take-off weight and maximum load. The Il-76M is the oldest variant to remain in use. The Il-76T was introduced as a civil derivative of the Il-76M, with military equipment removed and the payload increased again.

Two Il-76Ms taking part in the Victory Day Parade on 7 May 2021 over Moscow. Russian jet engines tend to be smokier than their Western counterparts and the Il-76's D-30s are no exception to this general rule.

Military production

Around 170 Il-76M/T aircraft were completed up to 1981, after which the Tashkent plant switched to the military Il-76MD and civil Il-76TD (D for *Dalniy*, long-range) with a reinforced wing structure and maximum take-off weight increased further. The lengthened Il-76MF first flew in August 1995, but never entered large-scale production. With a 6.6m (21ft) fuselage stretch, the Il-76MF (for fuselage) had capacity for 186 paratroopers or 305 troops in double-deck configuration. Powered by PS-90A76 turbofans, the maximum take-off weight increased to 210,000kg (46,2970lb).

Il-76M

One of the Russian Air Force's active fleet of Il-76Ms, RF-86842 is finished in the scheme typical of the type. The Il-76 has proven a highly successful aircraft; rugged, versatile and profoundly useful for both military and civil applications.

Ilyushin Il-76MD Candid
Weight (maximum take-off): 190,000kg (418,878lb)
Dimensions: Length 46.59m (152ft 10in), Wingspan 50.5m (165ft 8in), Height 14.76m (48ft 5in)
Powerplant: Four Aviadvigatel/Perm D-30KP series 2 (*izdeliye* 53) turbofans, each rated at 117.68kN (26,455lbf)
Maximum speed: 750-780km/h (466-485mph)
Range: 7800km (4847 miles)
Ceiling: 12,000m (39,370ft)
Crew: 5
Payload: 48,000kg (105,822lb)
Armament: Two 30mm (1.2in) Gryazev-Shipunov GSh-23 twin-barrel Gast autocannon flexibly mounted in rear turret

Special modifications

As well as basic transport versions, multiple special modifications were completed for military and civilian use. These include the Il-76MD Skalpel-MT, military airborne hospital with medical modules in the cargo hold and the related Il-76TD-S (S for *Sanitarnyi*, medical) civil airborne hospital. The Il-76K, Il-76MDK and Il-76MDK-2 (K for Kosmos, space) are cosmonaut training aircraft used to provide simulated weightless conditions. The Il-76MD-PS (*Poiskovo-Spasatelnyi*, or Il-84), is a search and rescue aircraft, with 40 rescue paratroopers, plus a motorboat and life rafts for 1000 survivors; although this remained a prototype only. The Il-76P (Pozharnyi) firefighting version can carry 44,000kg (97,003lb) of extinguishing agent in two tanks in the cargo hold, and the Il-76LL (Letayushchaya Laboratoriya, flying test bed) is a dedicated testbeds for engine trials. While the Il-76MF proved a dead-end, its PS-90A76 turbofans and new avionics were later incorporated in an upgrade for the Russian military, producing the Il-76MD-90 featuring a range of 4500km (2796 miles) with maximum payload of 50,000kg (110,231lb), glass cockpit and a flight crew reduced from six to five. Although only one example was completed and first flown after the upgrade in December 2005, Tashkent produced seven new-build Il-76TD-90VD (VD for Volga-Dnepr) civilian versions for Volga-Dnepr.

In December 2006, the Russian government decided to launch Il-76 production in Russia, at the Aviastar-SP factory in Ulyanovsk. This would involve a new version, the Il-76MD-90A (*izdeliye* 476) with an all-new new wing construction, maximum take-off weight increased to 210,000kg (462,970lb) and maximum payload to 60,000kg (132,277lb), PS-90A-76 engines and new avionics suite including glass cockpit, new communication suite and a new self-protection suite with warning sensors, active infrared jammer and flare dispensers, and a crew of five.

The first flight of the Il-76MD-90A prototype at Ulyanovsk took place on 22 September 2012. The next two examples were delivered to Beriev in Taganrog for conversion as A-100 airborne early-warning and control aircraft.

Il-78 tanker

Work on a tanker version of the Il-76 began around 1970 but only became a realistic proposal once the Il-76MD was available. The prototype of the Il-78 tanker made its first flight on 26 June 1983 and the Il-78 was officially commissioned into service in June 1987. Meanwhile, the improved Il-78M, with increased weight, was first flown in March 1987. In the Il-78, the rear gunner's position is occupied by refuelling operator.

After the demise of the Soviet Union production slowed; the final total built in Tashkent is 944 airframes, including 52 Il-78 tankers and 30 A-50 airborne early warning aircraft. The new production line was launched at Ulyanovsk in 2012. Today, Russian Aerospace Forces have around 100 Il-76s and 15 Il-78s. The main version in service with the Russian Aerospace Forces is the Il-76MD, normally fitted with a rear turret with two twin-barrel GSh-23 cannon.

Civilian versions are utilized by other state operators in Russia, of which the largest is the Ministry of Interior. Other air forces operating Il-76s include Algeria, Angola, Armenia, China, India, Iran, Ukraine and Uzbekistan. In October 2012, a contract for 39 new Il-76MD-90As from Ulyanovsk was announced by the Russian Ministry of Defence. The first aircraft from this order flew in August 2015 and was handed over to the defence ministry in December 2015. Russian Aerospace Forces also plans to upgrade around 40 existing Il-76MD aircraft to Il-76MD-M standard.

The Tashkent factory built 46 tanker versions between 1983 and 1993 (32 standard Il-78s, 13 Il-78Ms and one Il-78E for Algeria); later, six Il-78MKI aircraft were completed for India in 2003 and 2004. The first five Il-78s were delivered to the Soviet Air Force in 1985. Today, Russian Aerospace Forces Il-78s primarily serve Tu-160 and Tu-95MS strategic bombers. Some Il-78s inherited by Ukraine were later used as transports with refuelling equipment removed. Ukraine also sold four Il-78s to Pakistan as the Il-78MP.

Along with the decision to resume production of Il-76MD-90A transport at Ulyanovsk, it was decided to develop the new Il-78M-90A tanker aircraft based on it, which is able to deliver 40,000kg (88,1845lb) of fuel to 5000km (3107 miles) and is convertible between transport and tanker. In 2018 it was announced that Russia would acquire 14 Il-78M-90As and that the existing Il-78 fleet planned would be upgraded to Il-78M2 standard.

Taking part in the 2015 flypast marking 70 years since the end of World War II in Europe, this Il-78 was simulating refuelling two Su-34 fighters as they passed over Red Square, Moscow.

The Il-76MD carries up to 48,000kg (105,822lbs) of load in a pressurized freight hold. The cargo compartment seats 167 troops, or 245 with a second deck installed, or 126 paratroopers. Military equipment can be airdropped from high or low altitude. In the Il-78, two fuel tanks are fitted in the cabin with a total capacity of 28,000kg (61,729lb). Fuel is transferred via three refuelling pods, two under the wing and one on the port side of the rear fuselage.

Ilyushin Il-78 Midas

Weight (maximum take-off): 210,000kg (462,971lb)

Dimensions: Length 46.59m (152ft 10in), Wingspan 50.5m (165ft 8in), Height 14.76m (48ft 5in)

Powerplant: Four Aviadvigatel/Perm D-30KP series 2 (*izdeliye* 53) turbofans, each rated at 117.68kN (26,455lbf)

Maximum speed: 850km/h (530mph)

Range: 7300km (4,500 miles)

Ceiling: 12,000m (39,370ft)

Crew: 6

Payload: 100,000kg (220462lb) payload of T-6 military jet fuel

Armament: N/A

Il-78

In July 1993 '34 Blue' arrived in dramatic style at the Royal International Air Tattoo at RAF Fairford in the UK, by appearing over the airfield whilst simulating refuelling a Tu-95MS 'Bear-S'. The pods containing the refuelling drogues are apparent under the wings and on the rear fuselage.

Ilyushin Il-112

The Il-112V has been developed as a new-generation tactical transport aircraft under LVTS programme, intended to replace the An-26 and some An-12s.

The Il-112 originated in 1994 when the Russian government approved a project for a regional passenger turboprop, although no funding was assigned. In March 2000, the then Russian Air Force announced a tender for the LVTS (Lyogkiy Voyenno-Transportnyi Samolyot, lightweight military transport aircraft). This called for the movement of a load of 6000kg (13,228lb) over 3000km (1864 miles) with alternative loads of 60 troops or 37 paratroopers.

In April 2003 the Il-112V (Voyennyi, military) project won the LVTS competition, defence ministry funding was approved and first flight planned for 2006. However, a lack of military interest led to delays amid concerns of cost overruns resulting in programme cancellation in May 2010.

However, in 2014 the Ministry of Defence signed a new contract with Ilyushin for the relaunched Il-112V project. The first flight was planned for June 2017 but actually occurred on 30 March 2019. Since then, the programme has faced more setbacks, including the underpowered Klimov TV7-117ST turboprop engines that have failed to keep pace with the increasing weight of the aircraft: from 19,700kg (12,241 miles) as planned in 2003 (when it won LVTS competition) to 21,000kg (46,297 miles). With these engines, the maximum payload is reduced to 5000kg (11,023lb) carried over a range of 1200km (696 miles).

Weight reduction

After the first flight, the aircraft underwent significant changes since the structure proved too heavy. Weight reduction work was completed in September 2020 but during a test flight in August 2021, the aircraft was destroyed in an accident at Kubinka, killing all three crew.

As of mid-2020, plans to buy 48 Il-112V aircraft were announced and the next two aircraft were ordered for state tests. A planned civil variant is the Il-112T (*Transportnyi*, freighter).

Ilyushin Il-112V

Weight (maximum take-off): 21,000kg (46,297lb)

Dimensions: Length 23.49m (77ft 1in), Wingspan 25.74m (84ft 5in), Height 8.9m (29ft 3in)

Powerplant: Two 2800ehp (2059ekW) Klimov TV7-117ST turboprops

Maximum speed: 500km/h (310mph)

Range: 4800km (2983 miles)

Crew: 2–3

Payloadt: Up to 44 passengers or 5000kg (11,023lb) cargo

The first prototype Il-112 was lost after an engine fire broke out in the starboard engine and the aircraft crashed four minutes after taking off from Kubinka. This is the first prototype in happier times, taking off on a test flight.

RECONNAISSANCE

Modern Russian reconnaissance platforms consist of such diverse types as the Beriev A-50 'Mainstay' AEW platform, the airliner-derived maritime patrol aircraft the Ilyushin Il-38 'May', which has been seeking out NATO assets at sea since the late 1960s, and the radiation-hardened Ilyushin Il-80 'Maxdome', intended to maintain command-and-control functions in the event of nuclear war.

This chapter includes the following aircraft:

- Beriev A-50 and A-100
- Ilyushin Il-20 and Il-22
- Ilyushin Il-38
- Ilyushin Il-80
- Ilyushin Il-82
- Tupolev Tu-142
- Tupolev Tu-214
- Beriev Be-12

Instantly recognisable by its radar rotodome atop the fuselage, the Beriev A-50 is an AEW derivative of the ubiquitous Il-76 transport. This view shows well the lateral fins attached to the undercarriage fairings unique to this variant.

Beriev A-50 and A-100 NATO reporting name: Mainstay

Developed from the Il-76 by Beriev, the Mainstay is a strategic airborne early warning and control (AEW&C) aircraft based on the Il-76MD (A-50) or Il-76MD-90A (A-100) platform.

Development of a new AEW&C aircraft began in the 1960s, with the MNIIP (later Vega) institute responsible for Shmel mission system. In 1969, Ilyushin offered its Il-70 design that combined the Shmel with an Il-76 transport. In 1973, the Soviet government ordered Beriev in Taganrog to develop the A-50 based on the Il-76MD, also with Shmel. The first prototype of the A-50 (*izdeliye* A, programme Titan) completed its maiden flight at Taganrog on 19 December 1978, with the mission system installed during 1979.

The first series A-50 aircraft was flown in July 1985. Compared to the Il-76, it features a rotating dome for the mission radar above the fuselage, no nose glazing, stub wings added to the undercarriage nacelles (A-50 only) and additional cooling intakes. The flight crew comprises two pilots, a navigator, engineer and communications operator. The mission crew adds a commander, three tracking operators, three guidance navigators and

Beriev A-50

Shown as it appeared circa 2005, '46 Red' wears the standard white and grey livery of the A-50 fleet. As well as the obvious rotodome, the A-50 differs from the Il-76 in featuring a nose refuelling probe and deletion of the tail turret.

three engineers. The E-821 Shmel (bumblebee, or *izdeliye* R) radar and mission system possesses a search range of 230km (143 miles) for a fighter-size target at low altitude, or 300-350km (186-217 miles) at high altitude. The system can track up to 45 targets simultaneously and guide up to 12 friendly fighters to targets simultaneously.

After three prototypes, series production of the A-50 took place in Tashkent from 1985 to 1993. In total 24 series aircraft are believed to have been completed. Service entry at Siauliai (Lithuania) occurred in June 1984. Since 1998, the aircraft have operated from Ivanovo, Russia; today, around nine are in service. A planned upgrade launched in 1984 as the A-50M (Titan-M or **izdeliye**

Beriev A-50

Weight (maximum take-off): 190,000kg (418,878lb)

Dimensions: Length 49.59m (162ft 8in), Wingspan 50.5m (165ft 8in), Height 14.76m (48ft 5in)

Powerplant: Four Aviadvigatel/Perm D-30KP series 2 turbofans, each rated at 117.68kN (26,455lbf)

Maximum speed: 810km/h (503mph)

Range: 5100km (3169 miles)

Ceiling: 12,200m (40,026ft)

Crew: 15

Armament: None

2A) with the E-821-2 Shmel-2 mission system and new PS-90A engines was cancelled at the end of the 1980s.

Design upgrade

The aircraft are now being upgraded by Beriev in Taganrog to A-50U (*izdeliye* AU, programme Titan-U) standard with the Shmel-M (*izdeliye* RM) mission system. This is a new computing system with digital signal processing and a radar reportedly capable of detecting up to 150 targets at a distance of 600km (373 miles) and with new LCD screens for mission operators. The A-50U completed state acceptance tests in November 2009 and the first series upgraded A-50U was returned to service in October 2011. Five upgraded A-50Us were delivered by December 2018.

An export version was developed for India, with Israeli mission systems, as the A-50EI (*izdeliye* AI), with a contract for three aircraft signed in October 2003 and delivered between 2009 and 2011. The airframes are made in Tashkent, adapted for installation by Beriev in Taganrog, then finally equipped by IAI and Elta in Israel. One A-50 was to be equipped with the Israeli Phalcon radar system and sold to China but was cancelled in 2000. China instead installed an indigenous mission system, serving as the prototype of the KJ-2000.

Beriev has also developed a civil variant as the *izdeliye* 976, which is used to monitor aircraft and missile test flights.

A-100 successor

Beriev was contracted in 2007 to develop the A-100 aircraft as a successor to the A-50, using the new Il-76MD-90A airframe. The Premier radar has active electronic scanning in elevation and mechanical scanning in azimuth. The aircraft also features new navigation and communication systems plus additional antennas likely for passive surveillance. The first aircraft for conversion arrived at Taganrog in November 2014, but the programme has been severely delayed and the first flight tests with radar operational took place only in 2022. New A-100s are built by Aviastar-SP in Ulyanovsk.

Successor to the A-50, the A-100 is virtually identical in external appearance. This is the as-yet unpainted prototype A-100 about to make its maiden flight on 18 November 2017.

Ilyushin Il-20 and Il-22

NATO reporting name: Coot-A (Il-20), Coot-B (Il-22)

Both based on the Il-18D airliner, the Il-20 is a radar reconnaissance and electronic intelligence (ELINT) platform, and the Il-22 is an airborne command post.

The prototype for the Il-20, converted by Ilyushin from an Il-18D, made its maiden flight on 21 March 1968 – tests with the side-looking radar fitted starting in 1968. Work on an airborne command post began in September 1965, with Ilyushin preparing a draft design for the Il-22 Bison in 1967 and the first of two prototypes making its maiden flight in 1970. Both aircraft have a flight crew of five (pilot, co-pilot, navigator, flight engineer and communication operator) plus a tactical crew of six.

Side-looking radar

The first Il-20 series aircraft were produced by Znamya Truda in Moscow in late 1969. Production continued in small batches until 1976 for a total of 18 aircraft. Currently around 12 examples are operated by Russian Aerospace Forces on behalf of the Main Intelligence Directorate (GRU) of the General Staff of the Russian Armed Forces. Two-aircraft detachments are operated from various bases to monitor areas of interest (Baltic, Sea of Japan, Mediterranean, Syria, Ukraine, etc.) The Il-20's primary sensor is the Igla side-looking radar. The high-resolution antenna is carried inside a cigar-shaped 9m (29ft) long fairing under the forward fuselage and there are two large housings on the sides of the forward fuselage for oblique photo cameras or optional video cameras. Two vertical photo cameras can also be installed in the bottom of the fuselage. The fuselage housings also contain antennas for the Romb-4 ELINT system. Additional Kvadrat-2 ELINT systems with antennas in small domes are fitted below the rear fuselage. A Vishnya signals intelligence (SIGINT) system to monitor radio transmissions, using trapezoid antennas, is mounted above the forward fuselage.

The Il-20MS (*izdeliye* 17MS) series upgrade for the Il-20 was developed under the *Retsenzent* (reviewer) programme launched in 2009, based around the Novella NV1.17 reconnaissance system. The first Il-20MS *Retsenzent* (converted from Il-20M) made its maiden flight in February 2014. Il-20MS aircraft feature different external antennas and new mission computers with a fairing for the new radar below the rear fuselage. There is an electronic support measures (ESM)

Ilyushin Il-20M
Weight (maximum take-off): 64,000kg (141,056lb)
Dimensions: Length 35.9 m (117ft 9in), Wingspan 37.42m (122ft 9in), Height 10.17m (33ft 4in)
Powerplant: Four 3,170ekW (4,250ehp) Progress Ivchenko AI-20M series 4 turboprops
Maximum speed: 660km/h (410mph)
Range: 6000km (3728 miles)
Ceiling: 10,000m (32,808ft)
Crew: 11
Armament: None

Il-20M

This 'Coot' is shown as it appeared on 2 June 2020 when it was intercepted by Lithuania-based RAF Typhoon FGR.4s tasked with NATO's Baltic Air Policing mission. Il-20s regularly fly intelligence gathering flights between mainland Russia and the Baltic enclave of Kaliningrad.

system with antennas around the fuselage. Three cylindrical antennas are located on the upper fuselage, likely including satellite communications.

One Il-20M was mistakenly shot down in September 2018 by Syrian air defences; all 15 personnel on board were killed. Il-20 variants include: the Il-20RT (retranslator), used by Russian Space Forces for collecting and transmitting data on spacecraft and ICBM launches that features a long canoe fairing above the centre fuselage and no reconnaissance sensors; the unique Il-20M Anagramma fitted with large box-type antennas on the sides of the rear fuselage; perhaps ESM gear; and the Il-20M Monitor, used as part of an Earth survey system together with satellites with large cylindrical antennas on top of the fuselage, radar below the rear fuselage and an electro-optical turret below the forward fuselage.

Radio-relay system

Production of the Il-22 amounted to 31: nine Il-22s between 1976 and 1979 and 22 Il-22M Zebras, the radio-relay version with a shorter under-fuselage pod, between 1979 and 1985. Currently, Russian Aerospace Forces has around 15 Il-22s, most of them upgraded to Il-22M11-RT Sokol standard (RT for Re-Translator, relay) with the Sokol radio-relay system, converted from 1998, with the narrow cylindrical under-fuselage pod removed and featuring three hockey-stick antennas on the fuselage.

The Il-22 fleet also includes one Il-22K, with Sokol-K radio-relay system, and three Il-22PP *Porubshchik* (wood stealer), SIGINT and standoff jamming aircraft intended to jam enemy airborne early warning and control systems and air defence radars with four large dielectric fairings arranged in pairs on either sides of forward and rear fuselage and chaff/flare launchers for self-defence. Most are based at Chkalovsky outside Moscow.

Il-22M-11

One of the Russian Air Force's Il-22M-11 'Zebra' radio-relay aircraft in its Aeroflot derived colour scheme. These aircraft were active in the 2022 Russian invasion of Ukraine but the failure to neutralise Ukrainian air defences have required them to operate at a significant stand-off distance, limiting their effectiveness.

Ilyushin Il-38 NATO reporting name: May

A shore-based anti-submarine warfare (ASW) aircraft, with secondary maritime patrol and surveillance roles, the Il-38 is based on the Il-18 passenger aircraft.

Featuring the same wing, empennage, engines, landing gear and cockpit as the Il-18 airliner, the Il-38 programme was launched in 1960 with the aim of creating an ASW aircraft able to patrol for three hours at 2200km (1367 miles) from its base. The prototype made its maiden flight on 28 September 1961. Series production was undertaken by Znamya Truda (now the RAC MiG) factory in Moscow.

The first production Il-38 was completed in December 1967 and the last of 65 production aircraft was completed in 1972. The first aircraft entered service with the Soviet Navy evaluation centre at Mykolayiv, Ukraine, and from 1968 the Il-38 was fielded by operational regiments in the Northern and Pacific Fleets. Operated by a flight crew of five (two pilots, navigator, flight engineer and communications operator), plus a tactical crew of three (commander and two operators), the Il-38 utilized the Berkut ASW system controlled by a digital computer, managing a radio sonobuoy subsystem, search/attack radar (NATO codename Wet Eye) plus an indication and data input subsystem. The system allows semi-automatic and automatic cruise flight and patrolling in the operational area, radar search for surface targets (including search for submarine periscopes), automatic laying of a barrier of sonobuoys and tracking of buoys and processing data for attack. The passive omnidirectional buoy can detect a submarine at up to 2km (1 mile) and is deployed immediately before attack for exact positioning of the target. The aircraft mounts a magnetic anomaly detector (MAD) in a long 'sting' aft of the empennage.

Maiden flight

The prototype of the improved Il-38N was first flown in April 2001 but the programme was then suspended and work focused on the Sea Dragon for export. In its final form, the Novella-P-38 mission system was based on the Sea Dragon but with additional capabilities. Based around a computerized operational centre with two operator stations, each of which has two colour LCD displays and a control panel, the system combines

Ilyushin Il-38
Weight (maximum take-off): 66,000kg (145,505lb)
Dimensions: Length 40.14m (131ft 8in), Wingspan 37.42m (122ft 9in), Height 10.17m (33ft 4in)
Powerplant: Four 4250ehp (3,169ekW) Ivchenko Progress AI-20M turboprops
Maximum speed: 650km/h (404mph)
Range: 6500km (4039 miles)
Ceiling: 11,000m (36,089ft)
Crew: 7–8
Armament: Up to 8400kg (18,519lb) weapons and stores

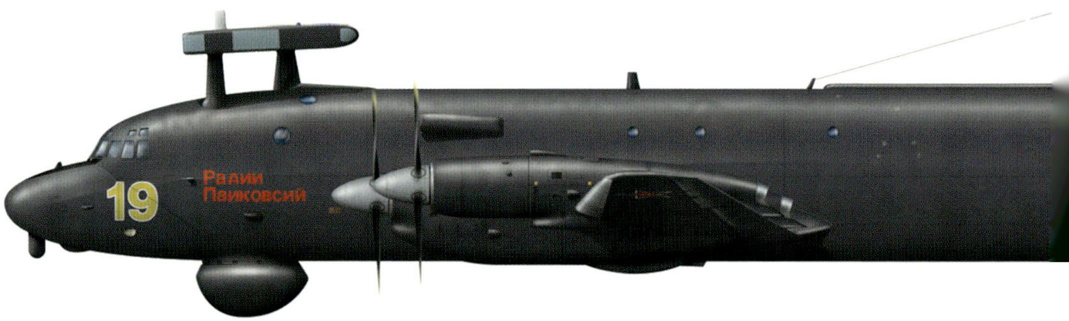

radar for detecting aerial and surface targets, a radio sonobuoy system with new buoys, improved MAD, an electro-optical turret (with TV, infrared imaging, laser rangefinder and target tracking sensors) and electronic support measures (ESM) sensors. The first fully-equipped Il-38N completed state evaluation in 2010 and was first fielded by the Northern Fleet in March 2012. In the same year the Russian Ministry of Defence signed a contract with Ilyushin for the upgrade of five aircraft to Il-38N standard. The first of these flew in April 2014 and was handed over to the Navy in July 2014. Another contract was issued in 2015 for the upgrade of five more aircraft.

Currently, the Russian Navy has 22 Il-38s at Severomorsk-1 in the Northern Fleet, and Yelizovo and Nikolayevka in the Pacific Fleet. Around 20 more aircraft are stored, some of which could be returned to service. The only export operator is India, which acquired five aircraft in 1977 with the downgraded Berkut-E system. All five Indian Il-38N aircraft are now upgraded to the Il-38SD with the Sea Dragon mission system.

Taxiing in after a patrol in July 2019, this Il-18N was operating from St Petersburg, the pylon-mounted boxlike structure houses ELINT equipment.

Rocket-propelled torpedoes

The Il-38 carries a maximum of 8400kg (18,518lb) of stores in two fuselage bays with a typical hunter/killer load of 154 sonobuoys, plus two torpedoes (including rocket-propelled torpedoes), or 10 depth charges or one nuclear depth charge. The upgraded Il-38N adds guided depth charges (up to 20) and improved rocket-propelled torpedoes. Indian Il-38SD can carry Kh-35E (AS-20 Kayak) anti-ship missiles on two underwing launchers.

Il-38N

Il-38s were originally painted in Aeroflot colours in an attempt to hide their true nature. Today they are finished in a more conventional overall grey. The Novella radar is housed in the ventral radome, with MAD equipment housed in the tail 'sting'.

Ilyushin Il-80 NATO reporting name: Maxdome

A strategic airborne command post serving the General Staff of the Russian Armed Forces, the Il-80 is intended to provide survivable nuclear command and control functions for the country's highest authorities in wartime.

Designed by Ilyushin in Moscow, the Il-80 was converted from the Il-86 wide-body airliner and operated by a flight crew of four plus an undisclosed number of operational crew. Hardened against the effects of nuclear explosion, the Il-80 also features additional fuselage fuel tanks and provision for inflight refuelling.

Series manufacture of the Il-86 and conversion to Il-80 was undertaken by Voronezh Aircraft Production Association (VASO) in Voronezh, Russia.

Submarine communication

First flown on 29 May 1985 without equipment fitted, the initial fully equipped Il-80 began flight tests in 1987. By 1990, it was taking part in exercises simulating the launch of ICBMs from the aircraft for the first time. Four Il-86 airliners were converted into Il-80 airborne command posts and originally operated with an Aeroflot colour scheme and titles. Based at Chkalovsky outside Moscow, together with the Il-82 radio-relay aircraft, the Il-80 is part of the Russian nuclear forces command system, known as 65s Zveno-S. The aircraft possesses a comprehensive communication suite including a large canoe-shaped fairing on the forward fuselage for satellite communication antennas. Fairings are mounted in front of the tailfin and under the rear part of the fuselage and the Fregat very low frequency (VLF) communication system is carried to communicate with submerged submarines. This uses a long trailing wire antenna released from a pod on the port side of the rear fuselage, with the cable extending several kilometres.

Of the four completed Il-80s, at least two have been upgraded since 2008 with the latest Zveno-2S system. Upgraded aircraft feature new antennas on the fuselage, including around a dozen very small blade-shaped antennas and a new hockey-stick antenna.

Il-80

The three Il-80s that reportedly remain operational are painted in standard Aeroflot livery but are not civil aircraft and are operated by the Russian Aerospace Forces. A derivative of the Il-96-400 is intended to replace the Il-80s.

Ilyushin Il-80

Weight (maximum take-off): 215,000kg (473,993lb)

Dimensions: Length 59.54m (195ft 4in), Wingspan 48.06m (157ft 8in), Height: 15.81m (51ft 10in)

Powerplant: Four Kuznetsov/Samara NK-86 turbofans, each rated at 127.48kN (28,660lbf)

Maximum speed: 850km/h (528mph)

Range: 7000km (4350 miles)

Ceiling: 11,000m (36,089ft)

Crew: Unknown

Armament: None

Ilyushin Il-82

The Il-82 is a strategic radio-relay aircraft that is designed to be used in conjunction with the Il-80 airborne command post.

Il-82

Confusingly, beneath the Aeroflot logo, this Il-82 displays the designation Il-76VKP. This was the designation by which it was known by its design team and Il-82 is the service designation of the two airframes.

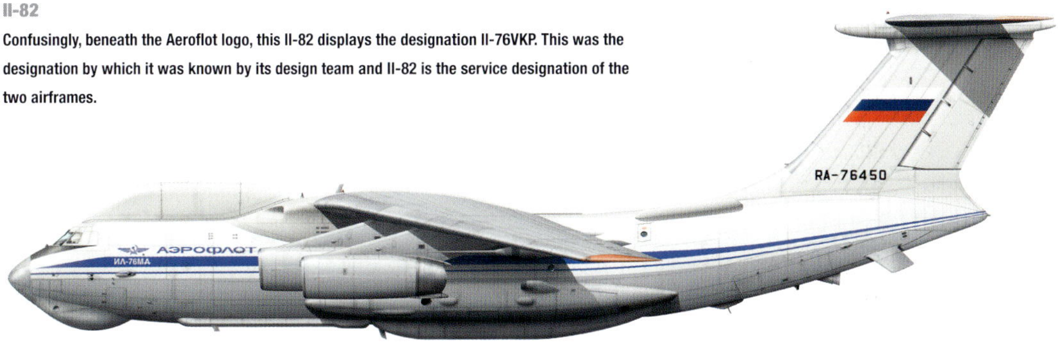

Developed and converted from an Il-76MD transport by Beriev in Taganrog, the maiden flight of the first Il-82 took place on 29 April 1987. In the event of a major war, the aircraft is designed to serve the General Staff of the Russian Armed Forces, operating in concert with the Il-80 airborne command post as part of wartime nuclear command and control.

Only two aircraft were converted and both are based at Chkalovsky airfield outside Moscow alongside the Il-80 fleet, as part of the 8th Aviation Division of Special Purpose to which these aircraft are assigned. Maintenance is carried out by Beriev, which also upgraded one of the aircraft in 2008, although the scope of the changes is unclear.

AI-24 turbines

The radio-relay system includes a large canoe-shaped fairing above the forward fuselage, containing satellite communication aerials. A trailing wire antenna for very low frequency (VLF) radio communication with submerged submarines is released from below the rear fuselage, with a stabilizing drogue at the tip.

Streamlined containers, with aerials at the front, are carried below the wingtips and numerous other longitudinal or sword-type aerials are fitted around the fuselage.

Il-86 derivative

Like the Beriev A-50, similarly derived from the Il-86, the nose glazing around the navigator's compartment has been deleted along with the crew entry door on the portside forward fuselage. Two AI-24 turbine engines are installed to supply power for onboard equipment and are mounted in the lengthened undercarriage nacelles. There has been a suggestion that a new radio-relay aircraft may be developed on the basis of the Il-76MD-90A but this remains unconfirmed.

Ilyushin Il-82

Weight (maximum take-off): 190,000kg (418,878lb)

Dimensions: Length 46.59m (152ft 10in), Wingspan 50.5m (165ft 8in), Height 14.76m (48ft 5in)

Powerplant: Four Aviadvigatel/Perm D-30KP series 2 (*izdeliye* 53) turbofans, each rated at 117.68kN (26,455lbf)

Maximum speed: 750–780km/h (466–485mph)

Range: 7800km (4847 miles)

Ceiling: 12,000m (39,370ft)

Crew: Unknown

Armament: None

Tupolev Tu-142

NATO reporting name: Bear-F (Tu-142MK and MZ) and Bear-J (Tu-142MR)

In 1963 the Soviet government called for Tupolev to develop a long-range ASW version of the Tu-95 strategic bomber able to patrol and destroy enemy submarines at 4000km (2485 miles) from its base.

Originally designated Tu-95PLO (*Protivo-Lodochnoy Oborony*, anti-submarine defence), by the time the aircraft made its maiden flight in Kuibyshev (later Samara) on 18 June 1968 the designation had changed to Tu-142 *izdeliye* VP, known to NATO as Bear-F Mod 1. It was operated by a normal crew of 10 (two pilots, two navigators, navigator/weapon system operator, two radio sonobuoy system operators, communications operator, onboard technician and rear gunner). Series production began in Samara the same year, the first 18 aircraft being fitted with the Berkut-95 mission system derived from the Berkut system used in the Il-38.

In 1971, production of Tu-142 switched to Taganrog, where a first example was flown in November 1975. Taganrog began to produce the upgraded Tu-142M (*izdeliye* VPM), Bear-F Mod 2 version with adaptations for the new Korshun ASW system,

Taken from a shadowing US aircraft, this Tu-142MR was photographed on one of its long range patrols in 1990. The longevity of service of the Tu-142 reflects that of the Tu-95 from which it derives.

Tupolev Tu-142MK

Weight (maximum take-off): 185,000kg (407,885lb)

Dimensions: Length 53.09m (174ft 2in), Wingspan 50.04m (164ft 2in), Height 13.62m (44ft 8in)

Powerplant: Four 11,185ekW (15,000ehp) Kuznetsov NK-12MV turboprops

Maximum speed: 855km/h (531mph)

Range: 12,000km (7456 miles)

Ceiling: 12,000m (39,370ft)

Crew: 11–13

Armament: Up to 9000kg (19842lb) of weapons and stores

An Indian Navy Tu-142MK-E. The Indian Navy flew the Tu-142 for 29 years, INAS (Indian Naval Air Squadron) 312 operating eight ex-Soviet aircraft between 1988 and 2017, when the Tupolevs were replaced with the Boeing P-8 Poseidon. IN-317 was photographed in February 2007.

but this was not yet ready and the Tu-142M was instead completed in 1975 and 1976 with the Berkut-95. In 1977, the Tu-142MK (*izdeliye* VPMK), Bear-F Mod 3, entered production with the Korshun-K search/attack system, controlled by a computer subsystem that brings together radar, radio sonobuoy subsystem with passive omnidirectional, passive directional, passive/active omnidirectional sonobuoys and tactical-data presentation. The aircraft also mounted a magnetic anomaly detector (MAD) on top of the tailfin. This variant was commissioned into service in November 1980. The final production version was Tu-142MZ (*izdeliye* VPMZ), Bear-F Mod 4 tested in prototype form in April 1985 and manufactured in 1993 and 1994 with a Korshun-N search/attack system with an upgraded computer subsystem, radar, and radio sonobuoy system, with new sonobuoys better able to detect quiet modern submarines. Both Tu-142MK and MZ can carry up to 9000kg (19,841lb) of stores in two fuselage weapon bays. A typical payload includes three torpedoes and 135 sonobuoys of different types.

The Tu-142MR (*izdeliye* VPMR) strategic radio-relay aircraft was authorized by the Soviet government in 1972 to provide a communications link between Russian national command authorities and commanders of ballistic missile submarines. First flown on 23 March 1978 with service entry in 1982, the Tu-142MR carries the Oryol communication relay system, with a suite of HF radios and satellite communications to connect ground-based or airborne command posts, very-low frequency (VLF) radio to connect submerged submarines, a data encrypting system and a relay system. Features include a trailing wire antenna installed in the forward bomb bay and deployed via a ventral pod, a satellite communications dome on the upper forward fuselage and a forward-pointing pod on the tailfin.

A crew of nine is carried (two pilots, navigator, onboard engineer, four system operators and rear gunner). The Tu-142MRM is an upgrade for theTu-142MR, with new communications equipment compatible with the latest Russian submarines.

Electronic Support Measures

Over 100 Tu-142s of all versions were completed. Currently, the Russian Navy has two Tu-142 squadrons, with Tu-142MK and Tu-142MZ versions, based at Kipelovo (Fedotovo) with the Northern Fleet, and at Mongokhto with the Pacific Fleet, for a total of around 16 airworthy aircraft. In addition, the Russian Navy operates around eight Tu-142MR/MRM aircraft. The Indian Navy operated eight Tu-142ME (*izdeliye* VPMK-E), Bear-F Mod 3, an export version with a downgraded Korshun-K-E mission system and different equipment including Israeli EL/M-2022A radar and electronic intelligence (ELINT) equipment and Indian electronic support measures (ESM). Indian aircraft operated from Dabolim from 1988 until 2017, when they were retired and replaced by the Boeing P-8I.

Tupolev Tu-214

The Tu-214R is a reconnaissance aircraft to supplement the Il-20 based on the Tu-214 airliner, with a flight crew of three (two pilots and a flight engineer), plus undisclosed mission crew.

The Tu-204 passenger aircraft first flew on 2 January 1989. The Tu-214 version has increased take-off weight and first flew on 21 March 1996. Series production of the Tu-204 was launched at the Aviastar-SP plant in Ulyanovsk, where 52 were built before production of the Tu-204 ceased and the Tu-214 began to be manufactured at the KAZ plant in Kazan.

Military role

Limited commercial orders led to the plans to develop the Tu-214 for military use, including reconnaissance, command post, radio relay, escort jamming and maritime patrol. In 2002, the Russian Armed Forces' Main Intelligence Directorate (GRU) ordered Tupolev and KAZ to begin research and development leading to two Tu-214R aircraft, or *izdeliye* 411.

Operational debut

The first Tu-214R took to the air on 24 December 2009 and the type had begun operational missions by December 2012. The Tu-214R utilizes a multi-band radar system with a side-looking radar, with flat antennas on the forward fuselage sides and a 360-degree surveillance radar in a radome under the rear fuselage.

SIGINT capability

A long, cigar-shaped fairing under the forward fuselage contains the electro-optical system. In addition, The Tu-214R has fairings on the fuselage sides, a cruciform antenna and two further radomes below the fuselage, likely for signals intelligence (SIGINT) system. Data can be transmitted using a communications suite, which includes dorsal satellite communication antennas.

Tu-214PU-SBUS

Details of the Tu-214's electronic, optical and radar equipment remain classified. The aircraft has been deployed over Syria as well as participating in the Open Skies 'mutual aerial observation' initiative until US withdrawal from the treaty in 2020 prompted Russia's exit in December 2021.

Tupolev Tu-214

Weight (maximum take-off): 110,750kg (244,162lb)
Dimensions: Length 46.14m (151ft 5in), Wingspan 41.8m (137ft 2in), Height 13.89m (45ft 7in)
Powerplant: Two Aviadvigatel PS-90A turbofans, each rated at 158.28kN (35,583lbf)
Maximum speed: 900km/h (559mph)
Range: 8000km (4971 miles)
Ceiling: 12,100m (39,700ft)
Crew: Unknown
Armament: None

Beriev Be-12 NATO reporting name: Mail

Designed and built by Beriev in Taganrog, this short-range anti-submarine warfare (ASW) and search-and-rescue (SAR) amphibian remains in service in Crimea.

The Soviet government ordered development of what became Be-12 (*izdeliye Ye*) in March 1956, the maiden flight taking place at Taganrog on 18 October 1960. Production aircraft had engines moved above the wings, raising the propellers clear of the water, plus radar in the nose. As well as the standard ASW version, the SAR Be-12PS (*izdeliye 3Ye*) was ordered with a capacity for 13 survivors and featuring medical and rescue equipment including a motor dinghy and starboard fuselage hatch. Between 1963 and 1973, a total of 140 series aircraft were built, 130 of them the ASW version capable of carrying nuclear depth charges, torpedoes, depth charges, mines, signal bombs and 10 SAR versions.

The first Be-12s were delivered to the Soviet Navy in 1964, the initial unit being the evaluation centre at Nikolayev, Ukraine. Operational ASW units were assigned to the Black Sea Fleet, Pacific Fleet, Northern Fleet and Baltic Fleet. Several civil conversions were completed in the early 1990s, including the firefighting Be-12P (*Pozharnyi*, *izdeliye YeP*) and the Be-12NKh (*Narodno-Khoziaystvennyi*, economic) for transport.

Engine overhaul

During the 1970s, 27 aircraft were upgraded to the Be-12N version, with a Nartsiss-12 targeting computer, a new sonobuoy receiver and new directional RPG-2 buoys as well as improved versions of MAD and radar. After the collapse of the Soviet Union, 55 aircraft remained in Russia and 14 in Ukraine. Today, a handful of Russian Navy aircraft remain active at Kacha air base, Crimea. The Ukrainian Navy has two aircraft in Mykolaiv, although their current status is unclear. In 2012, Russia conducted major overhauls on their Be-12s, including engine overhaul and new propellers, the aircraft returning to service between 2014 and 2015.

Beriev Be-12PS

Weight (maximum take-off): 36,000kg (79,366lb)

Dimensions: Length 30.11m (98ft 9in), Wingspan 29.84m (97ft 11in), Height 9.1m (29ft 10in)

Powerplant: Two 3863ekW (5,180ehp) Progress AI-20D series 4 turboprops

Maximum speed: 530km/h (329mph)

Range: 3600km (2237 miles)

Ceiling: 12,100m (39,698ft)

Crew: 4

Armament: None

'28 Yellow' was the first Be-12 to be overhauled at the EARZ plant in Yevpatoria. It was redelivered to the Black Sea Fleet in December 2016 and is seen here in July 2017.

Ту-134А

TRAINERS

Russia has historically relied on the aircraft produced by other Warsaw Pact nations for its training needs, notably the Aero L-39, which still serves in large numbers but is slated to be replaced by the Yakovlev Yak-130. The need for a low-cost basic trainer may have been met by the Yakovlev Yak-152 and crew training for larger types is handled by the now quite elderly Tupolev Tu-134.

This chapter includes the following aircraft:

- Tupolev Tu-134
- Yakovlev Yak-130
- Yakovlev Yak-152

Seen taxiing at Chkalovsky air base in 2010, this immaculate Tu-134A-3 executive staff transport was built in 1981, suffered a right-hand undercarriage collapse in 2007 but was repaired and reportedly remains on the active fleet.

Tupolev Tu-134 NATO reporting name: Crusty

A trainer for bomber pilots and navigators, the Tupolev Tu-134 was based on the Tu-134 regional commercial aircraft and had a flight crew of four – two pilots, navigator and flight engineer.

Design of the Tu-134 was inspired by the visit of Soviet leader Nikita Khrushchev to France in 1960, where he flew in a Caravelle airliner. He then ordered Andrei Tupolev to relocate the engines of the Tu-124 airliner to the rear fuselage. On 29 July 1963 the prototype Tu-134 (Tu-124A) made its maiden flight in Kharkiv. The aircraft entered scheduled service with Aeroflot in September 1967. The major passenger versions comprised the Tu-134 with 72 passengers, the Tu-134A with fuselage stretch and 76 passengers, and the Tu-134B for 80 passengers.

Production of the Tu-134 was undertaken in Kharkiv, Ukraine, between 1966 and 1984 and a total of 852 aircraft were built, of which 199 were military trainer versions.

The first Tu-134Sh (*Shturmanskiy*, for navigators) navigator/bombardier trainer flew in Kharkiv on 12 February 1971; 90 aircraft of this variant were built and used for training navigators of bomber aircraft and for bombing practice. Similar to the Tu-134A passenger version but with an enlarged radome under the glazed nose and bomb racks under the wing centre section, the cabin had 13 workstations for trainee navigators, the two at the front having a bombsight and a radar sight respectively. Two bomb racks, each for four 120kg (264lb) practice bombs, were fitted, later replaced with two bomb racks for six 50kg (110lb) practice bombs each.

Training mission

This was followed by the Tu-134UB-L (*Uchebno-Boyevoy dla Lyotchikov*, combat trainer for pilots), Crusty-B, for training Tu-22M3 and Tu-160 pilots that featured the nose of the Tu-22M 'grafted' onto a Tu-134B airframe. The passenger cabin has seats for 12 trainees who took turns occupying the co-pilot's seat during a training mission. The first Tu-134UBL bomber pilot trainer flew on 17 March 1981 in Kharkiv and 109 were constructed. A mid-life upgrade of the Tu-134UB-L resulted in the Tu-134UB-KM for training Tu-22M weapon system operators; two aircraft were converted in 1996 and 1999 at Pushkin. The nose contained the PNA radar from the Tu-22M3. Equipped with the optical bombsight from the Tu-22M, bomb racks were fitted under the centre section of the fuselage. Equipment was also fitted to simulate Kh-22M anti-ship missile launch procedures.

Today, around a dozen examples of the Tu-134Sh, Tu-134UB-L and Tu-134UB-KM trainer aircraft are operated by the navigator school in Chelyabinsk, a pilot school in Balashov, plus operational training centres in Ryazan and Yeysk.

VIP transporter

In the early 2000s, at least seven Tu-134UB-L aircraft were converted to VIP transport versions in Minsk, Belarus. These were designated Tu-134B3-M and carried 10-15 passengers. As well as the dedicated trainer versions and the Tu-134B3-M conversions, the Russian military also made use of the Tu-134K military staff version of 1970, with VIP compartment and an additional door in the rear fuselage; in total around 180 Tu-134K and Tu-134AK versions were built, with different cabin configurations for 26 to 47 passengers and various communications suites. In the mid-1980s, Tu-134AK aircraft began to be converted to Tu-134A/A-3 Balkany (Balkans) executive staff transport standard, with around 40 completed in Minsk. The Balkany, popularly known as the 'Salon' (lounge), featured a typical cabin configuration for 26 or 31 passengers including a VIP compartment in front, communication system room and standard 'economy' compartment in the rear. The HF radio communications system featured a long antenna 'sting' protruding from the rear fuselage.

Tupolev Tu-134Sh Crusty
Weight (maximum take-off): 47,000kg (103,617lb)
Dimensions: Length 37.05m (121ft 7in), Wingspan 29.01m (95ft 2in), Height 9.14m (30ft)
Powerplant: Two Aviadvigatel D-30 series II turbofans, each rated at 68.0kN (15,287lbf)
Maximum speed: 860km/h (534mph)
Range: 3400km (2113 miles)
Ceiling: 12,100m (39,700ft)
Crew: 4 (plus 13 trainees)
Armament: Up to 600kg (1323lb) of practice bombs

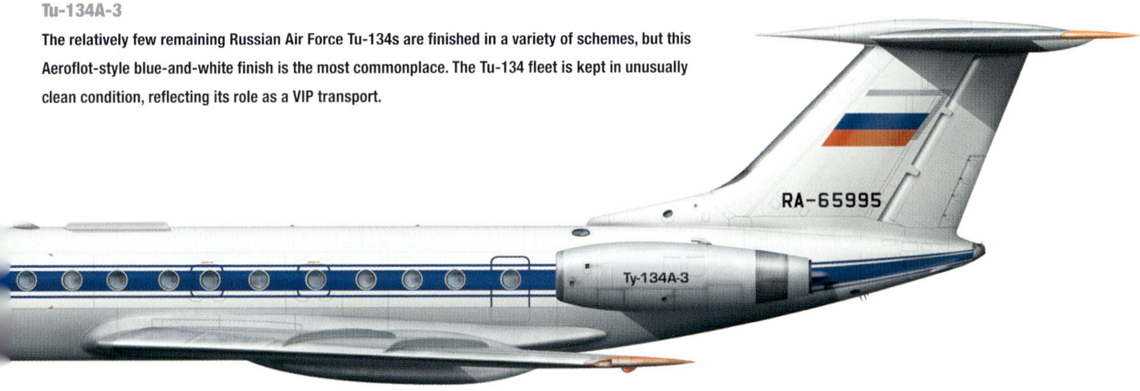

Tu-134A-3

The relatively few remaining Russian Air Force Tu-134s are finished in a variety of schemes, but this Aeroflot-style blue-and-white finish is the most commonplace. The Tu-134 fleet is kept in unusually clean condition, reflecting its role as a VIP transport.

Yakovlev Yak-130 NATO reporting name: Mitten

An advanced trainer and lead-in fighter trainer (LIFT), the Yak-130 supplements the L-39C Albatros and two-seat versions of combat aircraft in service with Russian Aerospace Forces.

In June 1990, the Soviet Union launched a competition for a new advanced jet trainer, with the Sukhoi S-54, Myasishchev M-200, Mikoyan 821 (later designated MiG-AT) and Yakovlev Yak-UTS (later designated Yak-130) designs all entered as responses.

By 1992, the competition had narrowed to the Yakovlev and MiG designs, but the final choice was postponed and prototypes of both the Yak-130 and MiG-AT were tested. The prototype Yak-130D made its maiden flight on 25 April 1996 and subsequently flew a total of 450 test flights until 2002.

Yakovlev and Aermacchi of Italy had cooperated on the AEM/Yak-130 project since 1993, leading to some aerodynamic redesigns plus a digital fly-by-wire flight control system and glass cockpit. However, the cooperation ceased due to disagreements between the two companies and their joint work ended in December 1999, with Aermacchi (later Leonardo) pursuing what became the M-346 Master.

'134 White' carries a pair of PTB-450 external tanks on the innermost pylons, along with two B-13L rocket pods carrying five 122mm (4.8in) unguided rockets each, and two R-73 air to air missiles on the outermost pylons.

Test flights

In March 2002, the then Russian Air Force approved the Yak-130, formally selecting it over the MiG-AT. Three aircraft were completed for tests, the first of these flying in April 2004. The second flew in April 2005 and the third in March 2006. The third was lost due to a fly-by-wire system failure in July 2006. Factory trials were completed in 2005 but the 2006 crash led to more changes (mainly to the flight control system). In late 2007, the first stage of state tests was completed and work could begin on the first 12 production aircraft. In December 2009, the Yak-130 completed state evaluation, clearing it for military use. Production aircraft can carry up to 3000kg (6614lb) of weapons and stores on six underwing hardpoints, plus one centreline hardpoint and two wingtip stations. Weapons include the R-73 (AA-11 Archer) infrared-guided air-to-air missiles, Kh-25M (AS-10 Karen) air-to-surface missiles, KAB-500L/Kr guided bombs, freefall bombs, rockets, gun pods or fuel tanks. The centreline station can accommodate a twin-barrel 23mm (0.9in) NSPU-130 gun pod.

Currently, Russian Aerospace Forces operate 109 Yak-103s, with four more having been lost in accidents. It was planned to equip combat units with the Yak-130 as a cheaper trainer to 'download' hours from fighter aircraft and maintain pilot proficiency, but this has apparently proven so far too expensive for such additional orders.

Pilot training

The production order for 12 Yak-130s was placed with the Sokol factory at Nizhny Novgorod in May 2005 and the first of these flew in May 2009. Four aircraft were sent to the test centre at Lipetsk, the remaining eight went to the training air base at Borisoglebsk in 2011, with student pilot training on the type commencing in 2013.

In 2004, the Irkut Corporation bought Yakovlev and after the four pre-series and 12 initial production aircraft were produced in Nizhny Novgorod, all other Yak-130s have been manufactured in Irkutsk. In December 2011, the Russian Ministry of Defence ordered 55 aircraft, the first of which was delivered to Borisoglebsk in October 2012; then, in December 2013, another 12 aircraft were ordered for a new aerobatic display team. The first export order came from Algeria for 16 aircraft in March 2006. Subsequent orders followed from Belarus (12), Bangladesh (16), Laos (four), Myanmar (20) and Vietnam (six).

Planned variants include the Yak-130M, a modernized version with a new mission computer, new head-up display (HUD) and a laser rangefinder in the nose, and the LUS (*Logkiy Udarnyi Samolyot*, lightweight strike aircraft): a combat version with new central computer, electro-optical targeting/navigation suite and a nose radar. Planned new armament includes air-to-ground guided missiles, laser-guided bombs, anti-ship missiles and beyond-visual range air-to-air missiles.

Yakovlev Yak-130 Mitten

Weight (maximum take-off): 10,290kg (22,679lb)
Dimensions: Length 11.49m (37 ft 9in), Wingspan 9.84m (32ft 3in), Height 4.76m (15ft 7in)
Powerplant: Two Ukrainian Progress/Zaporizhzhya AI-222-25 turbofans, each rated at 24.5kN (5510lbf)
Maximum speed: 1060km/h (659 miles)
Range: 1600km (994 miles)
Ceiling: 12,500m (41,013ft)
Crew: 2
Armament: Up to 3000kg (6614lb) weapons and stores

Yak-130

Appearing at the Kubinka Air display in August 2021, '30 Red' is finished in three-tone blue-grey camouflage and equipped with a pair of external tanks but no armament on its pylons.

Yakovlev Yak-152

Yakovlev's piston-engine trainer is intended for civil and military use, and series manufacture of the Yak-152 is planned at the Irkutsk Aircraft Plant, but the aircraft remains at the prototype stage.

Yak-152

Coded '01', this is the first prototype of a Yak-152 as it appeared at the MAKS 2017 show. It has since been joined by a second prototype, but the programme's future remains uncertain.

The Yak-152 programme to develop a trainer for *ab initio* instruction and flight screening has had several false-starts. Initially, the then Russian Air Force announced a tender for a successor to the Yak-52 in June 2001 to feature tandem ejection seats for student and instructor and the ability to carry gun pods, bombs and rockets on two underwing pylons. At this time, the Yak-152 was rejected in favour of the rival Su-49 in November 2001, but work never began in earnest.

In 2007, the programme for a primary trainer was relaunched by the Russian Ministry of Defence as the UTK PNP (*Uchebno-Trenirovochnyi Kompleks Pervonachalnoi Podgotovki*, training complex of initial training), but again yielded no results. At this point the rival to the Yak-152 was the Myasishchev M-107.

Little Bird

Yakovlev had already signed an agreement in June 2006 with Hongdu of China, stating that the firm would use the Yak-152 design for its CJ-7 trainer, also known as the L-7. The first prototype was completed in China by late 2010. More recently, Russia has resurrected the UTK PNP programme under the codename *Ptichka*-VVS (Little Bird, Air Force).

The design and mock-up of the Yak-152 was approved by the Russian Ministry of Defence in September 2014. However, the original plan to start series manufacture and deliveries in 2016 came to nothing. The Russian Ministry of Defence announced plans in September 2015 to order around 150 Yak-152s and the prototype Yak-152 finally made its maiden flight on 29 September 2016. Currently, two prototypes have been completed but their future is uncertain, and plans to extend the service life of Russian Aerospace Force's L-39C Albatros jet trainers suggests the Yak-152 programme may again be on hold, perhaps indefinitely.

Yakovlev Yak-152

Weight (maximum take-off): 1,490kg (3,285lb)
Dimensions: Length: 7.72m (25ft 4in), Wingspan: 8.82m (28ft 11in), Height: 2.47m (8 ft 1in)
Powerplant: One 370kW (500hp) RED Aircraft A03 V-12 diesel piston engine
Maximum speed: 500km/h (310mph)
Range: 1300km (808 miles)
Ceiling: 4000m (13,123ft)
Crew: 2
Armament: None

Index

References to illustration captions are in bold. References to photographs are in italics.

Picture Credits

Photographs
Alamy: 49 (Sergei Karpukhin/TASS), 62 (Stocktrek Images), 107 (Vega Radio Engineering Corporation/TASS)
Dreamstime: 6 (id1974), 8 (Upadek), 11 (Caspar107), 15 (Theflightvideo), 46 (id1974), 50 (Olonho), 72 (Kison1979), 90 (Fin808), 92 (Dragunov1981), 100 (Aarrows), 102 (Id1974), 117 (Sikaraha),
Dreamstime/Artyomanikeev: 34, 41, 57, 70, 104, 118, 121
Shutterstock: 7 (Dmitry Kosorukov), 19 (Media_works), 28 (Vladislav Sinelnikov), 67 & 111 (JetKat)
Shutterstock/Fasttailwind: 26, 27, 37, 48, 56
United Aircraft Corporation: 103
U.S. Air Force: 75 (TSGT Jim Varhegyi), 82 (MSGT Steven Turner), 95 (MSGT Christopher Hartman)
U.S. Navy: 65 (CWO Jeffrey Lund), 66 (MC2 Jason R Zalasky), 114
Alec Wilson: 115 (licensed under the Creative Commons Attribution-Share Alike 2.0 Generic Licence)

Artworks
Amber Books: 12/13, 20–22, 32/33, 34 bottom, 39, 42, 53, 54/55, 68, 76–79, 83–87, 90
Rolando Ugolini: 24–29, 88–89, 94–97, 100–123
Stuart Fowle: 30, 31, 34 top, 35, 36, 38/39, 40, 43–45
Teasel Studio: 10, 15–19, 22/23, 47, 48, 52/53, 56–61, 64–67, 69–74, 80–81, 91, 98, 99

"El evangelio de la prosperidad es un evangelio falso, y los predicadores de la prosperidad son sus falsos profetas. Este libro es así de claro —y persuasivo— pero es aún más que eso. Jones y Woodbridge han escrito un relato sencillo y cuidadoso de la nueva religión que se extiende por Asia, África y las Américas. Y han aportado una crítica justa, bíblica y penetrante. Me da tristeza decir que este libro es sumamente necesario, estoy agradecido de que ya esté disponible".

Mark Dever, pastor principal,
Iglesia Bautista Capitol Hill, Washington, DC

"El evangelio de la prosperidad no es ni prosperidad ni es el evangelio según la definición de las Escrituras. En pocas palabras, es una enseñanza falsa y una peligrosa herejía. Este libro, escrito por dos grandes estudiosos de la Biblia, investiga con esmero y exactitud, analiza y expone los errores teológicos y bíblicos que son una constante en ese movimiento. Esta es una obra importante y valiosa. Ruego a Dios en oración por su amplia distribución y lectura".

Daniel L. Akin, presidente,
Seminario Teológico Bautista del Sureste

"Jones y Woodbridge nos proporcionan en este libro la verdad acerca del evangelio de la prosperidad. Este breve estudio es muy claro y conciso, pues aporta los antecedentes históricos y la crítica bíblica de este movimiento. Su tratamiento es justo y equilibrado, ya que penetra hasta el núcleo central de la cuestión y revela el peligro del evangelio de la prosperidad. Recomiendo mucho este libro tan bien escrito".

Benjamin L. Merkle, profesor asociado,
Seminario Teológico Bautista del Sureste,
y editor de la serie de libros Preguntas y Respuestas

"Alabo a Dios por la publicación de este libro. Es un recurso confiable para todos los que quieren entender la naturaleza destructiva del movimiento del evangelio de la prosperidad. Jones y Woodbridge señalan cuidadosamente los principales errores bíblicos de ese movimiento y llaman la atención sobre la presencia de conceptos erróneos en la teología de la prosperidad. Al exponer estos errores, así como al interactuar con algunos de los principales defensores del movimiento de la prosperidad, Jones y Woodbridge han proporcionado un excelente recurso para ofrecer a los que han aceptado un evangelio falso, y para los pastores que desean proteger a sus rebaños de los lobos. Quiera Dios usar este libro para llevar a las personas al único evangelio verdadero, y al Salvador, que es nuestra herencia y tesoro".

Tony Merida, pastor/maestro, Iglesia Bautista del Templo,
y autor de *Faithful Preaching* y *Orphanology*

"Un enfoque excelente y equilibrado, el cual es rigurosamente bíblico y trata contenidos de mucha actualidad. Un recurso muy valioso para aquellos que deseen analizar este tema con precisión y claridad. Muy centrado en el evangelio".

J. D. Greear, pastor principal, Iglesia de la Cumbre,
y autor del libro *Breaking the Islam Code*